AF544733

U.S. CIVIL AIRCRAFT SERIES

VOLUME 9

U.S. CIVIL AIRCRAFT SERIES

VOLUME 9

(ATC 801 - ATC 817)

PLUS
GROUP 2 SECTION
MASTER INDEX

Joseph P. Juptner

TAB **AERO**
Division of McGraw-Hill, Inc.
Blue Ridge Summit, PA 17294-0850

This work is dedicated to the preservation and perpetuation of a fond memory for the men and the planes that made a future for our air industry. And, to help kindle a knowledge and awareness within us of our debt of gratitude we owe to the past. . . . . . .

Published by TAB Books
TAB Books is a division of McGraw-Hill, Inc.

hc 2 3 4 5 6 7 8 9 10 KGP/KGP 9 9 8 7 6 5

Library of Congress Catalog Card Number 62-15967

The Library of Congress has cataloged this serial publication as follows:

Joseph P. Juptner
U.S. Civil Aircraft, Vol. 9
Includes Master Index
Vol. for 1981
TL670 U15 629.13334 62-15967
ISBN 0-8168-9182-6

ACKNOWLEDGEMENTS

Any historian soon learns that in the process of digging for obscure facts and information he must oftentimes rely on the help of numerous people, unselfish and generous people many of whom were close to, or actually participated in, the various incidents or events that make up the segment of history recorded here; and, have been willing to give of their time and knowledge in behalf of this work. To these wonderful people I am greatly indebted, and I feel a heart-warming gratitude; it is only fitting then that I proclaim their identity.

My thanks to the American Aviation Historical Society; to Arthur P. Dowd; Richard S. Allen; Wm. E. Clark; to Peter M. Bowers noted author, historian & photographer; Ted Businger; George E. Cull aero. engr.; Ray Brandly of the National Waco Club; Bob Steele photographer; Bob Pickett of Cessna Aircraft; Chas. N. Trask (USAAF Ret.); Harry S. Gann of the McDonnell/Douglas Co.; Theron K. Rinehart of Fairchild-Republic; Ken M. Molson noted Canadian historian; Kenneth D. Wilson; Harvey H. Lippincott of United Technologies; Edward D. Williams of United Airlines; Wm. T. Larkins noted author, historian & photographer; John W. Underwood another noted author, historian & photographer; Gerald H. Balzer noted archivist; Gordon S. Williams of The Boeing Co. who always helps in the pinches; the indispensable Patty & Monty Groves of "Rare Birds"; Steve Fairfield of the "Hungry Eye"; and the many others who have done their bit.

FOREWORD

It is almost with a sigh of relief that I present this final volume of *U. S. Civil Aircraft*; to continue the series any further would take a younger mind with a fresher outlook on all that was developing in the era after World War II. For a writer who was brought up in the early era of aviation--and who lived it practically every waking hour—it has been a joy to recount the development of the commercial airplane. I did my best to picture the different airplanes with their fascinating shapes and lyrical names such as "Swallow," "American Eagle," "Red Arrow," "Skyrocket," "Flamingo," and others—airplanes that were just as fascinating as their names imply. This was at a time, too, when airplanes had distinct personalities, were colored gaily with every hue of the rainbow, and even had smells that excited the senses. It is doubtful that the airplane was more personal then than it is today; but surely it was time happily spent, when flying was out of the question and we just puttered around the airplane and delved into its intriguing innards.

Of course, the behavior of some of these older airplanes left much to be desired; but at least one felt better with the thought that they were much better than what we had before. Because of the relentless drive in aviation development it is still hard to realize how much aviation changed in the quarter-century between 1925 and 1950. The wildest dreamer of 1925 surely could not have predicted what airplanes would be like in the fifties!

This book has now entered the era of the fifties, and what lies ahead in aviation is almost apart from what had taken place in the decades before. The individual drive and enthusiasm for all things aeronautical was all but gone, the "backyard manufacturer" was a thing of the past, every experiment had already been proven or disproven, and the "lone eagle" type of aviator had gone out of style. Airplanes began to take on a sort of sameness, and companies building airplanes for civil use could be counted on less than ten fingers. This is not to say that aviation had relaxed and gone sedate—because progress and development did continue—but it just didn't seem the same old game. An airplane was no longer the vision of one or two men; a single airplane and an enterprising pilot did not a flying service make; and the aviator himself was no longer considered a special kind of man.

Surely, much of what has been written in this series of books are feelings and nostalgic memories out of one man's past fortified by extensive research into the times. But after twenty years of dedication to this project I found myself looking for a likely cutoff point. It presented itself more or less by a whim of the government agency that instituted the "Approved Type Certificate" used until the end of World War II. For more than twenty years until that time all ATC approvals were in numerical order (listed in this volume to #817).

But after the war and down to the present, approvals have been awarded by district. There are now six dispensing districts in the United States, and their approvals are allocated with such legends as 2A1, 5A9, 6A4. Such a system does not lend itself to any sort of recording in chronological sequence, and the end of the numerical sequence was a logical place to end this series.

Joseph P. Juptner

TERMS

To make for better understanding of the various information contained herein, we should clarify a few points that might be in question. At the heading of each new chapter, the bracketed numerals under the ATC number, denote the first date of certification; any amendments made are noted in the text. Unless otherwise noted, the title photo of each chapter is an example of the model that bears that particular certificate number; any variants from this particular model, such as prototypes and special modifications, are identified. Normally accepted abbreviations and symbols are used in the listing of specifications and performance data. Unless otherwise noted, all maximum speed, cruising speed, and landing speed figures are based on sea level tests; this method of performance testing was largely the custom during this early period. Rate of climb figures are for first minute at sea level, and the altitude ceiling given is the so-called service ceiling. Cruising range in miles or hours duration is based on the engine's cruising r.p.m., but even at that, the range given must be considered as an average because of pilot's various throttle habits.

At the ending of each chapter, we show a listing of registered aircraft of a similar type; most of the listings show the complete production run of a particular type and this information we feel will be valuable to historians, photographers, and collectors in making correct identification of a certain aircraft by its registration number.

In each volume there are separate discussions on 100 certificated airplanes and we refer to these discussions as chapters, though they are not labeled as such; at the end of each chapter there is reference made to see the chapter for an ATC number pertaining to the next development of a certain type. As each volume contains discussions on 100 aircraft, it should be rather easy to pin-point the volume number for a chapter of discussion that would be numbered as A.T.C. #93, or perhaps A.T.C. #176, as an example. The use of such terms as "prop," "prop spinner," and "type," are normally used among aviation people and should present nc difficulty in interpreting the meaning.

TABLE OF CONTENTS

ATC #801
(12-21-48)
EMIGH "TROJAN," A-2

Emigh "Trojan" A-2 shows its husky profile.

Most all airplane designers and aircraft engineers usually know what they want in an airplane before they put pencil to paper. Usually these preconceived attributes will dictate, more or less, of how this airplane is going to look. Harold Emigh, after years of study, made plans for an airplane that would be utter simplicity both in substance and in form; this an airplane that could be built cheaper than those of the time, and built quickly in the least amount of man-hours. He succeeded admirably in this quest, to be sure, but the lines necessary for simplicity surely were not the lines of charm and beauty. As a consequence, Emigh's little "Rocket/Trojan" lacked the subtle curves of grace, the esthetics of pleasing proportion, and that magnetic appeal of form that one is drawn to when seeing an airplane for the first time. This squat, dumpy-looking airplane that Emigh built could hardly provoke all that much interest, but in some ways he did have the right idea after all. Fashioned from easily-made stampings and large thick sheets of preformed metal, the "Trojan" was strong, carefree, and efficient; it also had a host of major components that were interchangeable with one another. Under the circumstances of mass production an airplane such as this could have been built cheaply by comparison, and built quickly; in this way an awful lot of airplanes could have been offered to the flying public at a price that many more could afford. Slowly nurtured into being after many years of setbacks and some tribulation, Emigh's little "Trojan" finally made the grade to production, but sadly enough, it was introduced in number at a time when the market for two-seated private-owner airplanes was "going to pot"; it was designed for a market that no longer existed in a sense, and that's the pity. But, there was some consolation—Emigh's "Trojan" was not the only one in this predicament. Most tallies show that a total of some 58 "Trojans" were built in all, and it is believed that no more than 8 or 10 are still flying. "Pop" Emigh did have the right idea, but it apparently came at the wrong time.

Harold E. Emigh learned to fly an "Eaglerock" (Combo-Wing) in 1929. He loved airplanes dearly and learned to understand them fully, but considered they were all too complicated. Having his own common-sense ideas on how an airplane ought to be built, Emigh made continuous studies of the situation in the normal course of things; he finally had one example practically finished in the mid-thirties, but a premature ill-timed test flight caused the loss of the airplane and death to the pilot. The Emigh-Gregg "Rocket" (NX-21963) of 1936 was another try at it and it showed great promise; it was still registered as experimental in 1939. However, the gathering clouds of war over Europe prompted Emigh to shelve the project, and to enter the nationwide CPTP program in 1941, here to help train a reserve of pilots for our military preparedness. When the CPTP training program finally phased out, Emigh flew with the Ferry Command, and of course, the "Rocket/Trojan" project still sat idle. After the war he settled in California and started the project again on a small airstrip in Bellflower; the prototype (NX-28390) made its first flight on 12-20-46, but hardly anyone noticed. Emigh also operated at Norwalk for a while where his family helped to build the airplanes, but that didn't work out too well either—it was tough to get started. Not one to give up easily, Emigh then moved to Douglas, Arizona in 1949 on the Douglas-Bisbee Airport where he received aid and encouragement to continue his development of the "Trojan." There was some production, but the whole project finally folded from lack of sales in 1950. Emigh stayed active in aviation with his mind still dreaming up novel concepts and an easier way to do things.

The Emigh "Trojan" model A-2 was an all-metal, low-winged cabin monoplane with side-by-side seating for two; the interior was rather chummy for 2 large people, but otherwise it was quite handy and comfortable. The "Trojan" was slung close to the ground in a seeming squat that some found unpleasant, but then others felt it looked especially docile and very friendly. From the stories bandied about in hangar-sessions it would seem that you either liked it, or you didn't. The "Trojan" has always been described as rugged, and as uncomplicated as a piece of farm machinery; that perhaps could have been its downfall. Most pilots wanted, and perhaps still want, their personal airplane to be of a feminine nature, and would tolerate them thus even if capricious, but the bandy-legged little "Trojan" just didn't measure up to this popular image. She was too much Plain-Jane and didn't generate any of that excitement that drew attention. As powered with the 85 h.p. Continental C-85-12F engine the hefty "Trojan" managed to be quite frisky in its deportment, and could hang in there with the best of its type. Its overall nature and characteristics were rather pleasant, albeit sometimes fidgety, and she did well out in the boon-docks, especially on the short grass-covered strips. Control was light and responsive, the airplane was very maneuverable, but she was emphatically restricted against intentional spins; symmetrical airfoils have very poor "stall" characteristics. Oh, it spun easy enough, and would recover fairly quick, but it wound up like a top and lost a lot of altitude even in normal recovery. This was probably more than the average low-time pilot would want to experience, even with a 6.3G positive safety factor. The "Trojan" was a toughie, it could play rough, and it wore well, but out of the 58 that were built there are less than a dozen still flying. Had Emigh (pronounced "Amy") put in a few more girlish curves, a little better proportion, and perhaps a little more caprice into his design it might have been an entirely different story. The

"Trojan" shows external ribs; wing was stout enough to walk on.

Simple lines of "Trojan" designed for ease of manufacture.

type certificate for the "Trojan" A-2 was issued 12-21-48 (for ser. #3 and up) and some 58 examples were built in all. Mfgd. first by the Emigh Aircraft Co. on Delpert Airpark in Bellflower, Calif., and then as the Emigh Aircraft Co. in Norwalk, Calif. About half or more of the production seems to have been done by the Emigh Trojan Aircraft Co. on the Douglas-Bisbee Airport in Douglas, Ariz. Harold E. "Pop" Emigh, Sr. was pres. & chf. engr.; C.M. Brakensiak was V.P.; Paul E. McKenna was gen. mgr.; Stewart D. Sippy was sales mgr.; Harold E. Emigh, Jr. was plant mgr.; & P.W. Corbett was chf. pilot.

Listed below are specifications and performance data for the "Trojan" A-2 as powered with Continental C-85-12F engine rated 85 h.p. at 2575 r.p.m. at SL; length overall 20'5"; height overall 6'5"; wing span 31'7"; wing chord tapered in planform & section; total wing area 156.8 sq. ft.; airfoil NACA-0012-63; wt. empty 874 lbs.; useful load 576 lbs.; payload with 28 gal. fuel 230 lbs. (1 pass & 60 lbs. bag.); gross wt. 1450 lbs.; max. level speed 126; red-lined at 156; cruising speed (.74 power) 115; landing (stall) speed 48; take-off run (over 50' barrier) 1460 ft.; climb 800 ft. first min. at SL; ser. ceiling 14,000 ft.; gas cap. 28 gal.; oil cap. (in sump) 4.5 qts.; cruising range (.74 power) using 5.4 gal. per hour was 550 miles; price first quoted was $2575, but average price later was $3295 at factory field. Also available with Continental C-90-12F engine rated 90 h.p. at 2475 r.p.m. for take-off; there was a slight increase in performance.

The fuselage framework was a simple all-metal (alum. alloy) semi-monocoque structure covered with stressed-skin panels of "Alclad" metal sheet; the preformed sheets ranged from .040 thick up forward to .032 in the aft section. The wing-walk was actually quite low to the ground providing easy walk-up and step-into entry into the no-frills cockpit. The cockpit enclosure was of sheet "Plexiglass" with quarter-windows behind for a look-see in back. Each canopy panel was slid up and over to the other side; only one side open at a time. An 8-11 cu. ft. baggage compartment for up to 100 lbs. was behind the seat-back. Dual stick-type controls were provided, and the dash-panel was sparse but neatly arranged. The cantilever wing was an all-metal (alum. alloy) two-spar structure of novel design that was covered with heavy-gauge, stressed-skin "Alclad" metal sheet; the metal skin was preformed and stretch-formed external ribs were riveted to its outer side to stiffen the structure and lessen the weight. The wing spars were more or less shear beams and spacers that provided attachment for the skin. The all-metal center-section was built into the fuselage underside providing space for the two fuel tanks (14 gal. each side with filler-neck), and attach points for the main landing rear struts; wing panels were bolted to the outer flanges. Long-span ailerons of metal framing were covered with (.025) "Alclad" metal sheet; each aileron was fitted with mass balance. There were no wing flaps, but the ailerons could be put into 3 deg. of "droop" for landings. The fixed tricycle landing gear of 79" tread used air-oil shock struts and 5.00x5 Goodyear wheels with hydraulic disc

Prototype "Trojan" shows slight difference from production models.

brakes; the 5.00x5 nosewheel was steerable with the rudder pedals. 1949-50 models had 6.00x6 wheels all round for better rough-field handling. The all-metal cantilever tail group was covered with (.025) "Alclad" metal sheet; the elevators had adj. trim tab. The rudder had fixed tab adj. on the ground only. The vertical and horizontal tail surfaces were interchangeable from side to side; in other words, the vertical fin could be one half of the horizontal stabilizer, and the rudder could be one of the elevators. The wing panels of a symmetrical section were also interchangeable from side to side, as were upper and lower skins of the fuselage. A McCauley fixed-pitch metal prop, normal set of engine & flight instruments, carburetor heater & air filter, fuel pump, fuel gauge, airspeed ind., compass, oil filter, navigation lights, 12V battery, & tie-down rings were std. equipment. Engine starter, generator & radio gear were optional. An engine starter & generator were std. equipment on the C-90-12F engine only.

It was noted that 18 of the "Trojan" were still on the U.S. register as of 3-31-80; the production prototype (ser. #3 as N-8300H) was still flying. Serial #3 was N-8300H and numbers ran in a continuous block to N-8356H which was ser. #59. The airplanes are pretty well scattered around the country with more early examples flying than the later ones. Mfgd. 1948-49-50.

Several of the "Trojan" have been rebuilt to fly again.

ATC #802
(9-23-48)
AERONCA "SEDAN", 15AC (S15AC)

Aeronca "Sedan" 15AC a roomy and economical four-seater.

The dedicated group at Aeronca Aircraft reasoned correctly that a reliable economy-class family-type airplane patterned somewhat after the popular "Chief" and "Champ" would have a good chance on the dwindling post-war market; this a market which had been swinging away from the all-too-numerous two-seaters, and looking towards the more useful "Family Flivver" four-seater. As a move to keep manufacturing costs to a minimum, a lesson learned in the past few years, a lot of the "Chief" and "Champ" components would be used for the new "Sedan" wherever possible; it is then no accident that the new airplane looked a little bit like either one of these when you first saw it. It is true also that the (15AC) "Sedan" was designed and built to a rock-bottom price tag (after all, price was perhaps the largest consideration in buying an airplane at this time), but it certainly didn't leave out the things that were most desirable in this type of utility airplane. Beside being of rugged construction with rather pleasing lines, it was utterly simple to operate and to maintain. It also had literally bags of room inside that could be used in so many different ways, and its built-in placid nature was something that everyone could enjoy. The 15AC had most of the things you would like about your own airplane. The "Sedan" first appeared in 1947 and after a little show-off time went into production in 1948. That Aeronca did hit upon the right formula for the times was reflected in the many orders and the hundreds of inquiries that began coming in. Before long the hustling "Sedan" could be seen all around the country; not making a big splash about it she served quietly from just about anywhere, faithfully doing just about anything that was asked of it. Truly, it was such a capable airplane that operators had a tendency to overwork it; but then, there were others owned by week-end pilots that were treated like one of the family. As deliveries of the "Sedan" continued, several were exported to neighboring countries where their utility was further exploited by bush-type operations on skis and floats; it was to become one of the darlings of the bush-pilot. The Aeronca "Sedan" was a comparatively cheap airplane; it was cheap to buy, that is, and cheap to maintain, but it was by no means a cheap "bucket-of-bolts." The "Sedan" was a sturdy first-line aircraft in a sensible design that has the operational record to prove it. According to various tallies perhaps over 400 units were built to 1950 when Aeronca Aircraft turned to (Korean War) war-work, and discontinued the manufacture of "Aeronca" airplanes. It was es-

Slope of "Sedan" fuselage provided extra lift.

timated recently that perhaps 200 or more of the "Sedan" (15AC) are still out there flying as good as they did some thirty or more years ago.

The stately Aeronca (15AC) "Sedan" was a mild-mannered, friendly-looking high winged cabin monoplane with more than enough room for four big people, and then some. It probably had more room inside than any other four-seater built at this time. Because it was designed for all-purpose service it operated equally as well on wheels, skis, or floats, and it wasn't too fussy about where it operated from. Several are even now still working over the difficult terrain, in all kinds of weather, over Alaska and Canada. That big efficient wing created a lot of lift and could pull a loaded "Sedan" up out of just about anywhere, even at the higher altitudes "where the air was thin." In fact, this was a "short-takeoff-or-landing" (STOL) airplane long before the word was even invented! The "Chief" and "Champ," because they were such popular airplanes, had built up quite a following of patrons "faithful to the marque"; many of these happy Aeronca-owners had moved up to owning a "Sedan." As powered with the smooth-running 6 cyl. Continental C-145 engine of 145 h.p. the (15AC) "Sedan" used every horsepower wisely and turned in an exceptional performance for its type of use; even the so-called difficult situations were taken in stride by a good pilot as an every-day event. Flight characteristics were especially pleasing, and tho' she took her time in getting around, like a ballerina in slow-time, her every action was predictable and precise. The "Sedan" was exceptionally stable and behaved well in "the stall," being just about as gentle and safe as an airplane can be. As one of the most useful airplanes of this time (late forties—early fifties) the "Sedan" sold well, was equally at home on big-city or small-town airports, and even "way out in the sticks." She hardly ever got fidgety, nor asked any favors; a placid airplane that took commands almost without question. Gathering from the lore built up over the years this was certainly quite an airplane, and there wasn't a mean streak in her anywhere. As the Aeronca "Sedan" she is fondly remembered by many, and is still being enjoyed. The type certificate for the 15AC (Sedan) was issued 9-23-48 and the seaplane version was approved in Oct. of that year; over 400 examples were mfgd. by the Aeronca Aircraft Corp. on the Municipal Airport in Middletown,

The "Sedan" was a capable seaplane.

O. John Lawler was pres.; S.F. Brady was V.P.; E.L. Sutherland was gen. mgr.; Floyd B. Simmen as sales mgr.; Raymond F. Hermes chf. engr.; Leon Wolfe proj. engr.; and Lou Wehrung as chf. pilot. Walton B. St. John later as sales mgr. Aeronca discontinued building airplanes in 1950 to do sub-contracting of military and space-work for other manufacturers.

Listed below are specifications and performance data for the Aeronca "Sedan" (15AC) as powered with 6 cyl. Continental C-145-2 engine rated 145 h.p. at 2700 r.p.m. at SL; length overall 25'3"; height overall 7'0"; wing span 37'6"; wing chord 66"; total wing area (including fuselage) 216.4 sq. ft.; airfoil NACA-4412; wt. empty 1150 lbs.; useful load 900 lbs.; payload with 36 gal. fuel 500 lbs. (3 pass. at 160 lbs. each & 20 lbs. bag.) payload with 26 gal. fuel 560 lbs. (3 pass. & 80 lbs. bag.); gross wt. 2050 lbs.; max. level speed 120; redlined at 139; cruising speed (2450 r.p.m.) 105 at 2000 ft.; landing (stall) speed 53; climb 650 ft. first min. at SL; take-off at gross wt. was less than 500 ft.; ser. ceiling 12,400 ft.; gas cap. 36 gal.; oil cap. (in sump) 8 qts.; cruising range (2450 r.p.m. at 5000 ft.) using 8.8 gal. per hour was 400 miles; price first quoted as $4795 lowered to $4395 in late 1948 and last pricing was $4848 with added equipment. The 6 cyl. Franklin 6A4-165-B3 engine was optional, and the Continental 0-300-A engine was also eligible. Seaplane gross wt. was 2100 lbs.

The fuselage framework was built up of welded chrome-moly steel tubing faired to shape with wooden formers and fairing strips, then fabric covered; the fuselage pinched down in the back to a slope that created extra lift. One large door (from the "Champ") and a convenient step on the right side offered entry into the spacious, sound-proofed cabin. All seats were quite comfortable and the rear seat could be quickly removed for stowage of extra baggage, or sporting gear; there was over 100 cu. ft. of space in the cabin with only the pilot aboard. A regular (7.2 cu. ft.) baggage compt. with allowance for up to 120 lbs. was behind the rear seat. Window area was ample, and the windshield sloped way back to the wing spar for visibility upward and to the sides. The dash panel had a glove compt. each side and one could be used for radio mount; a set of dual control wheels (from the "Chief") were provided. There were split-windows each side for extra ventilation, and the pilot's seat had two-way adjustment; front seats had folding backs. The slender semi-cantilever wing was a single-spar, all-metal (alum. alloy) structure covered with "Alclad" metal sheet; a single strut braced each side and the metal-framed ailerons were covered with fabric. A bladder-type (18 gal.) fuel tank was mounted in the root end of each wing-half; the wing tips were "washed out" to decrease their angle of attack and delay the "stall" of the outer wing. The landing gear of 84" tread was a simple tripod affair using rubber shock-cord (bungee) to snub the jolts; 6.00x6

Record-breaking "Sunkist Lady" flew from west coast to east coast and back without landing; view shows refueling method.

Some 15AC were fitted with "spray-rig."

wheels were fitted with 7.00x6 tires and Goodyear hydraulic disc brakes. The tail wheel was steerable. The 15AC was also eligible as a seaplane (S15AC) on Edo 89-2000 twin-float gear; as it came from the factory the seaplane was treated with corrosion-proofing and had an entry door each side. The fabric-covered tail group was built up of welded C/M steel tubing; both rudder and elevators were aerodynamically balanced. The elevators had adj. trim tab, but the rudder and one aileron had fixed tabs adj. on the ground only. A Sensenich wooden prop, electric engine starter, 12V battery, generator, normal set of engine & flight instruments, compass, air-speed ind., exhaust muffler, oil cooler, cabin heater, carburetor heater & air filter, toe-operated wheel brakes, parking brake, glove compts., fuel gauges, ash trays, cabin vents, navigation lights, landing light, stall-warning ind., two-tone paint, and tie-down rings were std. equipment. A McCauley fixed-pitch metal prop, "Aeromatic" vari-pitch prop, Bendix or G.E. radio gear, radio shielding, bonding, metal wheel pants, wheel fenders, Fram oil filter, prop spinner, an entry door on left side, rear seat cabin heater, seat covers, sun-visors, Edo floats, and skis were all optional. For 1949-50 the 15AC (Sedan) had a new paint scheme, improved interior styling, and was also approved with a spray unit to spray trees and crops! Standard two-tone colors were Red & Orange, or Green & White.

It was noted that 217 of the 15AC (Sedan) were still on the U.S. register as of 3-31-80; 15 were registered as the S15AC on floats as seaplane. Serial #2 (as N-1000H) the first production airplane was still flying as was ser. #5; the highest serial number registered was #561; 59 were registered in Alaksa & the rest were pretty well scattered around the country.

ATC #803
(3-15-49)
MOONEY "MITE," M-18-L (M-18-C)

M-18-L skoots above countryside; Peter Bowers flying.

The spiffy little Mooney "Mite," if one remembers back, bore the identifiable style and swagger of the designer, one Albert W. Mooney. Based somewhat on the PQ-8 and PQ-14 "drones" of World War 2 fame, the pinch-penny "Mite" was designed in 1946 to achieve the lowest cost transportation of any means then known; it was a "magic carpet" that could fly for an hour or more on a good deal less than a dollar's worth of gas! Designed first around the little 25 h.p. Crosley "Cobra" automobile engine the "Mite" promised to be the weekend pilot's dream, but its production was soon suspended while development went ahead to increase the engine's power rating; subsequently the use of this engine was dropped in favor of the popular 4 cyl. Lycoming 0-145 engine of 65 h.p. This engine change had some detrimental effects on the overall economy, but the performance gained was now akin to that of a small fighter-plane, and ex-fighter pilots took a fancy to it in droves. There was nothing around that could match the challenge, and the sheer fun, of a first time solo flight in the new (M-18-L) "Mite." Being a real sweet-flying airplane it generated much talk amongst the flying-folk everywhere, and it was certainly interesting enough to be covered well by the news media of that time. Delivery of the first M-18 "Mite" (with Crosley "Cobra" engine) came late in 1947, and as the first production model it reportedly went to a Mooney dealer in Calif.; he flew it home from Wichita (about 1200 miles) for less than 7 dollars! The M-18-L with Lycoming engine was being delivered from mid-1948; in the nearly 5 years it was produced at Wichita some 239 of the little "Mite" were rolled out for fly-aways, but ever-rising labor and production costs had driven up the price to where the "Mite's" major selling advantage was about gone. The last version of the "Mite" (M-18-C55) with 65 h.p. Continental engine, and several deluxe improvements, had to sell for nearly $3800 and that didn't help matters either. Seeking lower overhead expenses the Mooney operation moved to Kerrville, Texas in 1953 where some 50 or more or so of the "Mite" were built. But, it too became a losing proposition; the bulk of the aircraft market was now demanding a more comfortable airplane with more capacity and more utility. Oddly enough, the vivacious little "Mite" is much more appreciated nowadays than it actually was during the time it was being built. Now they say the "Mite" is back, only this time you have to build it yourself from plans and a kit based on original production blueprints. As a light-hearted airplane designed for owner-pilots who want sheer fun in their flying, with now and then a little high adventure at low cost, Mooney's little "Mite" was hard to beat.

An intriguing sight, the (M-18) Mooney "Mite" was a small, inexpensive low-wing cabin monoplane with just barely enough room for the seating of one; because of the chummy layout the pilot virtually became one with the airplane. Standing jauntily on its three-legged landing gear it seemed to beckon like a sorceress that could lead you up and astray from what was then known as every-day normal flying. Almost like having the fabled "seven-league boots" the wandering "Mite" has been to the lower tip of So. America and back, and as far north as the Arctic Circle; beside that it set many world records in speed, distance, and altitude for its class. Its primary reason for being was economy, flying on a very low budget (you could easily fly for as much as 300 hours including fuel, oil, maintenance, and insurance for just over $600), but the built-in exciting performance was the added bonus that came on to warm the heart of many a man. As first powered with the little Crosley "Cobra," a modified automobile engine of 25 h.p., it didn't have all that much performance to brag about, but the "Mite" really came alive as powered with the 65 h.p. Lycoming engine; performance with the 65 h.p. Continental engine was comparable. Slick as a whistle you might say, the "Mite" was then fast enough to scoot across a mile in less than 30 seconds, and it could do just about anything a pilot would care to try (except major aerobatics) with a flick of the stick. It didn't have hands-off stability, and it had to be tended to lovingly just about every minute, but it was absolutely viceless, and that was the charm of this little airplane; it was a sweet-flying airplane that you could really enjoy once you decided to go along and share the fun.

An M-18-C55 with Monty Groves in the cockpit.

Men show relative size of the "Mite."

As one owner-pilot so aptly put it "There was a delightful period in my life when I owned and flew a "Mooney Mite." That just about says it all! Nowadays if you should want an airplane of this type you'd have to build it yourself. About 80 of the M-18-L that were produced were limited to 780 lbs. gross weight, and then gross weight was increased to 850 lbs. in the new M-18-LA of which nearly 50 were built. The model M-18-C had 65 h.p. Continental engine at 850 lbs. gross wt. and about 120 of these were built; the M-18-C55 (last version) of which about 40 or so were built had a larger cockpit area and a bigger canopy—the basic price by then had gone to $3695 and up. The type certificate for the M-18-L was issued 3-15-49 for ser. #2 and up; the M-18 with "Cobra" engine was approved earlier, and the Continental-powered M-18-C was approved 4-11-50 for ser. #201 and up. Approval for the M-18-LA was issued 7-12-50 for ser. #100 thru #200, and approval for the M-18-C55 was issued 4-2-55 for ser. #323 and up. Some 290 or more examples in the "Mite" series were mfgd. by Mooney Aircraft, Inc. in Wichita, Kan. (formed in 1946) with the last batch (about 65) built in the plant at Kerrville, Tex. Chas. G. Yankey was pres.; Al W. Mooney was V.P., gen. mgr., & chf. engr.; Art B. Mooney was V.P. in chrg. of prod. & development; N.E. Miller was asst. engr.; W.W. "Bill" Taylor was sales mgr. & chf. pilot. Al Mooney left the company with brother Art in 1955, but before leaving he had designed the now-famous "Mark 20" (approved on TC 2A3) which has become the basis for a whole new line of "Mooney" airplanes. The M-18 with Crosley "Cobra" engine was deleted from ATC #803 on 4-28-50.

Listed below are specifications and performance data for the Mooney M-18-L as powered with 4 cyl. Lycoming 0-145-B2 engine rated 65 h.p. at 2550 r.p.m. at SL; length overall 17'7"; height overall (at rudder) 6'3"; wing span 26'11"; wing chord 56" at root tapering to 28" at tip;

M-18-C was the most popular "Mite."

total wing area 95.05 sq. ft.; airfoil NACA-0015 Mod. at root tapering to NACA-4410 at tip; wt. empty 500 lbs.; useful load 280 lbs.; payload with 12.8 gal. fuel was pilot at 165 lbs. & 30 lbs. bag.; gross wt. 780 lbs.; max. speed 138; cruising speed (2300 r.p.m.) 122 at 5000 ft.; landing speed (with flaps) 40; stall speed (no flaps) 45; normal takeoff run 525 ft. over 50' barrier; takeoff run (full flaps) 350 ft.; climb 1090 ft. first min. at SL; climb to 10,000 ft. was 12 mins.; landing run (over 50' barrier) 860 ft.; ser. ceiling 19,400 ft.; gas cap. 12.8 gal. (11 gal. usable); oil cap. 4 qts.; cruising range (.66 power at 10,000 ft.) using 3.5 gal. per hour was 360 miles; price for M-18-L was $2250 at factory. The M-18 with "Cobra" engine was listed for $1995 and the M-18-C55 built in Kerrville sold for $3695-3800. The M-18-L at 740 lbs. gross wt. had positive safety factor of 4.4G, and later versions had safety factor of 3.8G because of higher gross wt. The M-18-C and M-18-LA both eligible at 850 lbs. gross wt.

Specifications and data for model M-18-C as powered with Continental A-65-8 or A-65-12 engine rated 65 h.p. at 2300 r.p.m. at SL, same as above except as follows: length overall 17'9"; wt. empty 520 lbs. (Std) & 580 lbs. (Deluxe); useful load 330-270 lbs.; payload with 12 gal. fuel was pilot at 170 lbs. & 80-20 lbs. bag.; gross wt. 850 lbs.; max. speed 140; cruising speed (.68 power) 125 at 10,000 ft.; landing speed (with flaps) 43; stall speed (no flaps) 48; takeoff run over 50 ft. barrier 560 ft.; climb 1000 ft. first min. at SL; land over 50' barrier was 900 ft.; gas cap. 12 gal.; oil cap. (in sump) 4 qts.; cruising range (.68 power at 10,000 ft.) using 3.8 gal. per hour was 350 miles; price for M-18-C (Std.) was $2325 & $2545 (Deluxe) at factory. Specifications & data as listed here also typical of M-18-LA and M-18-C55.

The fuselage framework was a composite structure using welded C/M steel tubing for the center portion that was covered with metal panels; the rear portion was a plywood-covered, all-wood, semi-monocoque structure covered with a final layer of fabric. The transparent canopy slid back for entry into the cockpit from the wing-walk; the canopy could be fully open in flight for that open-air feeling. The canopy could also be opened just slightly for ventilation. The M-18-C55, as the cream of the crop, had a larger cockpit area that was sound-proofed with fiberglass, upholstered with imitation leather, and had a bigger canopy. Stick-type controls were provided as well as manual controls for landing gear retraction and lowering of the wing flaps. The fuel tank (variously of 8, 12, 13.5, &

"Mite" with Crosley "Cobra" engine; note belly-slung radiator.

16 gal. cap.) was high behind the pilot and a 3.6 cu. ft. baggage compt. was down underneath behind the seat. The cantilever wing (of high aspect-ratio) was a box-type, single-spar structure with a plywood leading edge, wooden ribs and trusses, & then covered with fabric; ailerons and the slotted hi-lift wing flaps were also covered with fabric. No flap deflection above 85 m.p.h. The "Mite" was also equipped with a version of the Simpli-Fly system (as used on the "Culver V") called Safe-Trim; the entire tail assembly moved in coordination with the wing flaps changing angle of horizontal stabilizer so that the airplane was constantly "in trim" as the wing flaps were cranked up or down. The retractable (hand-operated) tricycle landing gear of 71" tread used Mooney rubber-donut shock struts with 4.00x4 Mooney wheels fitted with toe-operated Mooney brakes; the nosewheel was steerable. A novel Wig-Wag system (very dramatic) was on the dash-panel to warn of "gear up" when the engine was throttled down; this a very effective reminder to lower wheels on landing. The plywood-covered, wood-framed tail group was also covered with a layer of fabric, and all movable controls were aerodynamically balanced; the elevators had adj. trim tab. The little "Mite" was the first "Mooney" with the "on backwards" tail. A Sensenich or Flottorp wooden prop, exhaust collector, carburetor heater & air-filter, normal set of engine & flight instruments, compass, airspeed ind., wiring for navig. lights, fuel gauge, & seat belt were std. equipment. The standard M-18-L color scheme was overall aluminum finish with a contrasting stripe. A full electric system, 12V battery, engine starter, generator, cabin heater, navig. lights, & prop spinner were std. equipment on the M-18-C

prop spinner were std. equipment on the M-18-C Deluxe & the M-18-C55. Optional eqpt. for most std. models at extra cost was carb. air-filter, cabin heater, navig. lights, ash tray, 6V or 12V battery, electrical system, starter, generator, battery, a landing light, shielding & radio gear; all this was provided on Deluxe models except the radio gear.

To celebrate his 25th anniversary (6-24-50) as an airplane designer, Al W. Mooney took off from Brownsville, Texas in an M-18-L with 45 gals. of fuel aboard, and flew to Watertown, S.D.; this a distance of 1312 miles in less than 11 hours to set an unofficial distance record for Category One (under 1102 lbs.) airplanes. Proving that the Mooney "Mite" altho' the smallest and cheapest airplane then in production, was one of the most efficient airplanes ever built. Mooney encouraged record-setting in the "Mite," so many records and record-attempts followed, some breaking his own record. Hoping to cash in on the "Mite's" ability an experimental version labeled the M-19, and referred to as the "Cub Killer," was demonstrated in 1951 with 90 h.p. as a miniature fighter-plane with twin .30 caliber machine guns and rocket launchers. It was a whale of an airplane for its purpose, but the government services which instigated this design, suddenly lost interest.

It was noted that 166 of the Mooney "Mites" were still on the U.S. register as of 3-31-80; 48 were the M-18-L, 21 were M-18-LA, 80 were the M-18-C, and 16 were the M-18-C55. Ser. #3, 5, 6, 7 were the earliest ser. nos. shown & #357 was the highest number. All were pretty well scattered around the country with perhaps the larger portion in the west; one was registered in Alaska.

ATC #804
(5-18-48)
LUSCOMBE "SEDAN," 11A

Luscombe (11-A) "Sedan" was a sturdy and very practical four-seater.

The Luscombe "Sedan" (11A) was a rather unusual airplane in many ways, but it still managed to look and act pretty much like a "Luscombe" despite the many obvious differences. Following the lead of several manufacturers who had gotten into the burgeoning four-seater market, Luscombe Airplanes designed the "Sedan" as a convertible work-horse airplane for such as the traveling salesman, the many-acre farmer, the doctor, the rancher, and even the so-called bush-pilot; that it was also an ideal family-type airplane was an extra bonus. The Model 11A (Sedan) program was initiated with gusto in June of 1946 and by November of that year the bare-bones prototype was already flying; several hundred hours of shake-down flying proved the design to be just exactly as planned—a thoroughly sound design. They at Luscombe Airport felt good about this airplane and spent a lot of money to develop it, paying much attention to useful features, and several eye-catching innovations. While one airplane was being tested for government approval, several others were on demonstration in various parts of the country to get a reaction; the reaction was enthusiastic and the ship proved itself adaptable to varied uses from all kinds of terrain. To further showcase the new "Sedan" a dressed-up version was sent to the "Flying Farmers" convention in Aug. of 1947 where it strutted its stuff and made a very good impression; most lookers agreed it was a lot of airplane for the $6500 price tag quoted, and the orders started coming in. And, because it was so versatile it began to appeal to a varied cross-section of users and operators.

The "Sedan" program for some reason was interrupted for a while, but it was finally approved on 5-18-48. Production was launched early in 1948 with the first delivery being on the 5th of June; despite the late start some 57 airplanes (11A) were delivered in the balance of that year. The "Sedan" fared quite well despite the strong competition from other types, because it did have a lot to offer by comparison, but the market at that time seemed to be fizzling and drying up fast; by year's end Luscombe Airplanes was left almost begging, and their "Silvaire" production had just about run out also. Had Don Luscombe still been with the company, 'twas said, he may have pulled the outfit out of its doldrums with something from his bag of tricks, but it actually was overspending for development and the subsequent financial difficulties that caused the operation to fold up. Luscombe Airplanes reluctantly went into bankruptcy in the Spring of 1949 and was taken over by nearby "Temco" (also in Dallas) who was then building the feisty little "Swift," and also had taken over manufacturing rights for the incomparable "Fairchild 24." Temco did build the two-seated "Silvaire" into 1950, and a pitiful few of the (11A) "Sedan"; ironically the four-seated "Sedan" had been left to die on the vine. The "Luscombe" company is now long gone, but there are perhaps nearly 50 of the "Sedan" out there either flying, or waiting to be restored. Perhaps the "Sedan" offered too much, and was built too far ahead of its time.

The Luscombe (11A) "Sedan" with its strong stance and broad shoulders was a distinctive-looking, high-winged cabin monoplane with just oodles of room for four; in fact, it had an exceptional amount of room for stretch-comfort. It also had visibility down the nose and all around that could hardly be matched in this type of airplane. Its utility was also enhanced because loading and unloading was easy, the seats were quickly removable, and space was then ample enough for even bulky items; this found favor with the farmers and the ranchers. Then too, simplicity was a big part of its makeup so that it would be easy to operate, and easy to maintain; another bonus was its ability to operate out of the smaller unimproved 'fields, or even a farmer's pasture. As powered with the husky 6 cyl. Continental E-165 engine of 165 h.p. the "Sedan" conducted herself in a very capable manner with a rather high performance by standards of the time. Flight characteristics were pleasant thruout the entire range of flight, and the ability was there for unusual circumstances. It had a high degree of stability and actually flew as well when loaded as it did empty; altho control was light, you could tell you were in a big machine. It was rather easy to fly and fun too, but she had just enough quirks to hold your interest. From accolades tossed its way, then and even now, one would think that the Luscombe "Sedan" was some kind of "Seventh Heaven"! To many it was because she didn't have a mean rivet in her whole frame. The "Sedan" instilled pride of ownership too because she was a little out of the ordinary, and you were always noticed wherever you went; it turns heads even today. Had the market been more favorable to the "Sedan" at the time it's a cinch-bet that many hundreds would have been built, and many more would now be flying. The type certificate for the (11A) "Sedan" was first issued on 5-18-48 (sometimes listed as 10-4-48) and 57 examples were mfgd. by the Luscombe Airplane Corp. at Dallas, Texas. One L.H.P. Klotz was pres.; Stewart L. Whitman was sec-treas.; O. W. Hoernig was sales mgr.; Eugene W. Norris was V.P. of engrg.; H. "Gus" Erickson was chf. engr.; "Herb" F. Kueck was asst. engr.; & Horace Scroggins was chf. pilot. "Herb" Kueck fashioned the aerodynamic design of the 11A, and "Gus" Erickson designed the structure. Temco Aircraft Corp. who had taken over was also in Dallas, Texas.

Listed below are specifications and performance data for the Luscombe "Sedan" (11A) as powered with 6 cyl. Continental E-165 engine rated 165 h.p. at 2050 r.p.m. at SL; length

"Sedan" performance was ideal for cross-country jaunts.

Businessmen found owning the "Sedan" very useful.

overall 23'6"; height overall 6'10"; wing span 38'0"; total wing area 165 sq. ft.; airfoil NACA-4412; wt. empty 1280 lbs.; useful load 1000 lbs.; payload with 40 gal. fuel 571 lbs. (3 pass. at 170 lbs. each & 61 lbs. bag.); max. bag to 100 lbs.; gross wt. 2280 lbs.; max. speed 140 plus; cruising speed (.74 power) 130; landing speed (with flaps) 55; stall speed (no flaps) 60; normal takeoff run 900 ft.; climb 900 ft. first min. at SL; normal landing run 1200 ft.; ser. ceiling 17,000 ft.; gas cap. 42 gal.; oil cap. (in sump) 10 qts.; cruising range (1850 r.p.m.) using 10 gal. per hour was 500 miles; price first quoted at $6500, soon raised to $6995 with standard equipment. The Continental E-185 engine was an optional installation for better performance.

The fuselage framework was an all-metal (alum. alloy) semi-monocoque structure covered with stressed-skin "Alclad" metal sheet. A large 30" wide door and a convenient step was on each side for easy entry into the sound-proofed cabin; the spacious interior was upholstered in maroon and tan broadcloth. Adjustable front seats were deeply padded over foam rubber, and the seat-backs folded forward; the view was exceptional down that slanted nose. The comfortable bench-type seat in back was removable for added space. A 6 cu. ft. baggage compt. was under the rear seat with allowance for up to 100 lbs.; with pilot only and all other seats removed there was 55 cu. ft. of space with allowance for up to 600 lbs. of tied-down cargo. The large dash-panel was neat, well arranged, and had a glove compt. each side; dual wheel-type controls were provided, but toe-operated brake pedals were on the left side only. Window area was exceptionally large for a panoramic view. There was a dual skylight above with draw-curtains to keep out the sun's glare, and the rear cabin window for 360 deg. vision was quite novel; windows in each door could be opened out for extra ventilation. A hat-shelf behind the rear seat, leather assist straps, ash tray & cigarette lighter at each seat, a map-case on each door, door locks, & ignition lock were some of the other conveniences. The semi-cantilever wing of a high aspect-ratio was an all-metal (alum. alloy) two-spar structure having only a few wing ribs, and then covered with heavy gauge stressed-skin "Alclad" metal sheet; the wing was braced each side by a single metal strut. Metal-framed ailerons and wing flaps were covered with ribbed metal skin; no flap deflection above 100 m.p.h. Wing flaps were operated by a hydraulic hand-pump to 30-40 degs.; wing flaps used only to steepen the glide path. The rudder and aileron controls were interconnected by a bungee-cord system for simple coordinated turns, but could be overridden if desired. A 21 gal. fuel tank was mounted in each wing root, with about 20 gals. usable. The cantilever "Silflex" landing gear of 100" tread was fitted to oil-spring shock struts mounted beneath the floor; 6.00x6 Goodyear wheels with 7.00x6 tires were fitted with Goodyear disc brakes. The spring-leaf tail skid was fitted with a Maule steerable tail wheel. The cantilever tail group was an all-metal (alum. alloy) structure covered with "clad" metal sheet; the aerodynamically balanced elevators had adj. trimming tab and were limited to 13 deg. of up-travel. This made it rather difficult to get the tail down in landing with no one in the rear seat. A Sensenich or

Hartzell wooden prop, engine starter, generator, 12V battery, carburetor heater & air-filter, oil cooler, normal set of engine & flight instruments, exhaust muffler, fuel gauges, compass, airspeed ind., cabin vents, cabin heater, parking brake, floor rug, sunshades, dome light, panel lights, navig. lights, prop spinner, cargo tie-down straps, tie-down rings, & all controls on ball bearings were generally as std. equipment. Landing lights, dual brake pedals, radio gear, fixed-pitch metal prop, or Hartzell two-position prop were some of the options available.

This approval for ser. #11-4 and up mfgd. prior to 10-29-56; approval expired as of that date. It is interesting to note that 36 of the (11A) "Sedan" were still listed on the U.S. register as of 3-31-80. Serial #11-5 was still flying in Mass. & ser. #11-198 was the highest serial number listed. Two of those listed in the register were Temco-built; one of these (#11-195) is still flying in Texas. Four were reported in Canada, but status is unknown The 1600B to 1690B block of registration numbers were allotted to the "Sedan" and various "Silvaire" that were built at that time.

Luscombe people proudly display price-tag.

Prototype 11-A (Sedan) on one of its early flights.

ATC #805
(8-26-48)
PIPER "VAGABOND DELUXE," PA-17

PA-17 (Vagabond) was a sporty addition to Piper line.

One of the most delightful of the 65 h.p. lightplanes of this period was the perky "Vagabond," a rather plain-looking little "cheapie" that fooled everybody and practically became America's sweetheart. A cheaper economy-version of this design had already been approved earlier by Piper as the PA-15, and its sales literally put the company back on its feet again. The new version of the "Vagabond" featured here was more or less a look-alike called the PA-17, as "Vagabond Trainer" or "Vagabond Deluxe." Actually, the 2 airplanes (PA-15 and PA-17) were very much alike, but the PA-17 was distinguished, or set apart, from the PA-15 by having what the other little airplane didn't have. This later version, the PA-17, was powered with the more expensive Continental A-65 engine, and it was fully equipped with all the things that had either been unavailable, or were offered only as options on the earlier economy model. Being the more sporty airplane of the two, the price-tag for the PA-17 was naturally higher, and slanted towards the owner pilot who could better afford to pay for the extras; nowadays one might say that $200 or so extra is no big deal, but 30 or more years ago it posed as quite a difference. Wm. T. "Bill" Piper was credited with saying "with this airplane we're going to find out if there still is an owner-pilot market out there." Well, there was, but just barely. From the visual standpoint the "Vagabond" (either one) looked something quite like a big model-airplane; it looked so petite and ordinary that it was downright cute. It definitely had that homely charm that made it fun to own. As time had later proven, the "Vagabond" was a delightful trouble-free airplane that had never heard a truly harsh word from anyone, and seemed to hum with its own happiness. The "Vagabond" was the first of the so-called "short-wing" Piper airplanes, and from it were developed such all time favorites as the enchanting "Clipper" and the "Pacer" series. It is also worth mentioning that like many another manufacturer Piper Aircraft during this time had developed a promising selection of diverse models slated for the post-war market. After all the hoopla had finally died down the post-war climate became very unfavorable to the promotion of new designs, and every one of them died on the vine you might say.

The stubby little Piper "Vagabond," in this case the PA-17, was a light short-coupled, high-winged cabin monoplane with side-by-side seating for two. It was not too big, not too small, and laid out in proportions that seemed just right for a rather cheap low-powered airplane. Because this new version was slanted towards owner-pilots with a little more spending-money in their wallet, it offered most of the niceties that made owning an airplane a little more pleasant. The interior was now finer, an optional extra door was a big convenience, the landing gear now used honest-to-goodness shock absorbers, and many more inexpensive options were available to dress her up in sporty fashion. Even this newer version (the PA-17), altho' not as Plain-Jane as the PA-15, still posed with an awkward grace as if shy, but once she was "fired up" and in the air it became a fun-partner that was easy to enjoy. As powered with the 65 h.p. Continental A-65 engine its overall performance was a lot better than one would expect, and she could even get a little saucy at times. Because of its short-couple, and its small wing span, direction changes and the roll-rate could be enough to quicken the pulse; controls responded so lightly and so crisply it always brought out a grin of disbelief in the first-timer. But then, the "Vagabond" was not always so saucy, she did have patience and dignity. It was also exceptionally stable for such a light airplane; it would "trim out" nicely for hands-off, but she didn't like to be unattended for very long. "Stalls" were rather gentle with quick recovery, but "intentional spins" were frowned upon because recovery required a certain expertise. Despite the modest price the "Vagabond" (PA-17) was no cheap "clunker"; it was a good, sturdy airplane that was sensitive to mood and it responded to the way it was handled. Perhaps that is why it was so enjoyable. As one owner-pilot put it "I had mine for more than 10 years, and it was the joy of my life"! An airplane has to earn praise such as this. The type certificate for the model PA-17 was issued 8-26-48 for ser. #17-1 and up; it is quite probable that well over 200 examples of this model were mfgd. by the Piper Aircraft Corp. at Lock Haven, Pa. Wm. T. Piper, Sr. was chrmn. of brd., pres., gen. mgr., & treas.; H. Van Bortel was V.P.; Walter Jamouneau was sec. & chf. engr.; John Hinson was sales mgr.; David Long was proj. engr.; & Frank Gibson was chf. pilot.

Listed below are specifications and performance data for the "Vagabond" model PA-17 as powered with Continental A-65-8 engine rated 65 h.p. at 2300 r.p.m. at SL; length overall 18'8"; height overall 72"; wing span 29'3"; wing chord 63"; total wing area 147.5 sq. ft.; airfoil USA-35B Mod.; wt. empty 650 lbs.; useful load 500 lbs.; payload with 12 gal. fuel 250 lbs. (1 pass. & 80 lbs. for bag & extras); bag. allow. to 40 lbs.; gross wt. 1150 lbs.; max. speed 102 at SL; cruising speed (.80 power) 92 at 2000 ft.; landing (stall) speed 48; takeoff run (over 50' barrier) 1570 ft.; climb 510 ft. first min. at SL; landing roll to stop (over 50' barrier) 1280 ft.; ser. ceiling 12,000 ft.; normal gas cap. 12 gal.; optional gas cap. 18 gal.; oil cap. (in sump) 4 qts.; cruising range (.80 power at 2000 ft.) using 4.3 gal. per hour was 250 miles; price $2195 at factory. The Continental A-65-8F engine was optional; this engine featured engine starter & generator.

A PA-17 rebuild with 85 h.p. and electrical system.

A converted "Vagabond" (PA-17) with tricycle gear.

The stubby fuselage framework was built up with mixed grades of welded steel tubing, scantily faired to shape, then fabric covered. As an added convenience there was an optional door each side for entry into this model. The cabin interior of 39 in. width had more comfortable seats, and was neatly upholstered with serviceable fabrics; the dash-panel was quite large and most everything was quite handy, but grouped toward the left side. Dual stick-type controls were std., but dual brake pedals were optional. A 12 gal. fuel tank was mounted up front behind the dash panel with a float-type gauge visible from the cockpit; a 6 gal. aux. fuel tank was optional. The side-windows slid open for ventilation, and a baggage hamper (9 cu. ft.) with allowance for 40 lbs. was behind the seat. A very efficient (stainless steel) exhaust muffler and cabin insulation kept the interior fairly quiet; the cabin was now fitted in more detail and there was a rug on the floor. The abbreviated wing framework in 2 panels was built up with extruded dural spar beams and formed metal wing ribs; the completed framework was covered with fabric. The long-span ailerons of the offset-hinge type were very effective because of their large area. The landing gear of 69 in. tread was a simple tripod affair using 2 spools of rubber shock-cord and 8.00x4 Goodrich wheels with hydraulic toe-brakes; the full-swivel tail wheel was steerable. The fabric-covered tail group was built up of welded steel tubing; the left elevator had adj. trim tab and the rudder had aerodynamic balance. The rudder & elevators were strong & very effective. A Sensenich wooden prop, carburetor heater & air filter, oil filter, normal set of engine & flight instruments, airspeed ind., exhaust muffler, fuel gauge, adj. cabin vents, compass, prop spinner, seat belts, & tie down rings were std. equipment. A McCauley fixed-pitch metal prop, wheel pants or wheel fenders, dual brake pedals, landing lights, skis, aux. fuel tank, cabin heater, & radio gear were optional.

It was noted that 131 of the PA-17 were still on the U.S. register as of 3-31-80. Serial #2, 3, and 4 were still flying and the highest ser. no. on the register was #17-215. All were pretty well scattered around the country.

ATC #806
(3-12-48)
DE HAVILAND-CANADA "BEAVER," DHC-2

The fabulous DHC "Beaver"; not all were kept so "dandy."

An in-depth story on the famous DeHaviland of Canada "Beaver" would be a rousing book in itself, so we must of necessity, touch only upon a few highlights of its romantic existence. Only a few times before had an airplane made such an impact on worldwide aviation. To begin with, DeHaviland of Canada was formed in 1928 as a service depot for the various "DH" equipment being used in Canada. In 1937 they began building complete airplanes to DeHaviland of England design; the DH-82 "Tiger Moth" trainer, which is now so popular in the U.S.A., was built up thru' World War 2 to handle the accelerated training program. A small amount (about 50) of the DH-83 "Fox Moth" were also built, and probably 1200 or so of the famous "Mosquito" fighter-bomber were built for the war effort. After hostilities ended the DHC-1 "Chipmunk" was the first post-war design built by DeHaviland of Canada, a training machine which first took to the air in 1946. After losing nearly all production of the "Chipmunk" to the parent company in England, the Canadian group turned its attention to designing the DHC-2 "Beaver," a workhorse, no-nonsense airplane that was fashioned to cope with the frontier's miseries. The prototype DHC-2 made its first flight on 8-19-47 in Toronto with Russell Bannock at the controls just one year after preliminary engineering had started. From early tests they at "DHC" estimated the "Beaver" would be a "smashing success," but to just what extent they surely could not then envision. It was to become Canada's most popular export.

The hardy "Beaver" was ever-popular because of its robust construction, its load-carrying abilities, and its knack of making an airport out of any little strip. Also, its utility on wheels, skis, or floats, and its unwavering reliability in the most difficult service was another bonus. Realizing that the "Beaver" was specifically designed to conquer the elements of any hostile continent, operators from all over the world, where conditions for operating an airplane were particularly hostile, were putting in orders; undeveloped areas just naturally became the "Beaver's" habitat. It is logical to assume that the "Beaver" would surely be used in Canada and Alaska first, the stomping grounds of bush-pilots for many years, but crated airplanes were soon going to the

"Beaver" as a seaplane; the bush-pilot's favorite.

remote regions of Malaya, Singapore, So. & Central Africa, the Persian Gulf, Iran, Australia, New Zealand, and thereabouts. Because its worldwide reputation was steadily mounting, its services were even sought here in the U.S.A. As the military L-20 the "Beaver" was evaluated by the U.S. Army Field Forces and the USAF at Fort Bragg, N.C. starting in the early part of 1951; the tests were a great success and the airplane was subsequently ordered in large numbers by special legislation. Eventually there were over 980 units in U.S. military service, the last being delivered in 1960, and many were used for the fracas in Korea & Viet Nam. When the shenanigans in Viet Nam wound down to a halt, thousands of surplus military airplanes were slated to go on the auction block. One of the first such sales took place in Arizona in 1976 where 104 of the "Beaver" (L-20A/U-6A) were up for grabs. News of the "Beaver" sale created a terrific amount of interest amongst civilian operators, and calls came in from as far away as Alaska and So. America. The U.S. government had paid some $99,000 per copy for the "Beaver" and they were still getting $20,000 to $30,000 for the somewhat shabby lot. The "Beaver" was used in nearly 65 different countries, an airplane that has been just about everywhere from pole to pole, and it is hard to predict when its service will end. Because they were always used in most difficult service many of the "Beaver" got pranged and bent up, that is understandable, but they were always fixed and put back into service.

The slab-sided DeHaviland (DHC-2) "Beaver" was a good-sized, high-winged cabin monoplane with normal seating arranged for a pilot and 6 or 7 passengers; with seats removed and pilot alone it could carry stacks of cargo and just about anything that would fit in it, or on it. It even had provisions for lashing a canoe, or a stack of lumber to its underside! Versatile enough to operate on wheels or skis, it was especially the bush-pilot's favorite on floats. Float landings could be made on grass too, and wheels could be swapped for the pontoons in less than half a day, or vice versa. The "Beaver" was actually not all that sophisticated but put together with quality workmanship, and much of the actual repairs needed from time to time could be done by the pilot in the field with only a simple tool kit. As powered with the tried-and-true 9 cyl. Pratt & Whitney "Wasp Jr." (R-985) radial engine of 400-450 h.p. the droning "Beaver" was no speed-ball, but its rib-tickling short-field performance was among the very best; it could

Hard-working military "Beaver" as the U-6.

operate in and out of places where a good many other airplanes feared to tread. Fashioned as a wayfarer to begin with it could be seen hard at work in the scorching heat, or the numbing cold, at sea level, or high up in the mountains. Because it was built more or less like a truck the "Beaver" flew heavy like a big airplane, but it was well-behaved, responded well, and tried to make the pilot's job as easy as possible, especially in a pinch. "Beaver" pilots jokingly called it ugly, because it really wasn't a raving beauty from any angle, but all will say to a man that it was an airplane easy to be fond of; it was a rough and tumble airplane (stressed to 3.5 G) and it respected a pilot's trust. The first "Beaver" had made its debut in 1947 and production lasted for 20 years into 1967; in that time more than 1600 had been built, and perhaps half of those are still out there flying. The fact that serial #1 was still flying regularly in 1969 is also quite unbelievable. The type certificate for the DHC-2 "Beaver" Mk. 1 was issued 3-12-48 and some 1630 examples were mfgd. all together by DeHaviland Aircraft of Canada Ltd. in Toronto (Downsview), Ontario. A "Turbo-Beaver" with turbine engine (turbo-prop) was introduced in 1963 and approved on 2-18-66, but it did not take the place of the standard piston-engined "Beaver." This approval for ser. #1 thru' #79, #81 and up; the Mk. 2 with Alvis Leonides 502/4 engine was approved 5-6-53 and only one was built. Revision of 10-28-60 noted that (Army) L-20A were also eligible when modified to conform to DHC-2 Mk.1 standards. Some L-20A were approved on

"Beaver" on amphibious floats with U.S. Border Patrol.

(Restricted Category) AR-33 issued 4-1-60 to Civil Air Patrol, Inc. of Anchorage, Alaska for special purpose of search & rescue, or pest control.

Listed below are specifications and performance data for the DHC-2 Mk. 1 "Beaver" as powered with P & W "Wasp Jr." SB-3 (R-985-AN) engines rated 400 h.p. at 2200 r.p.m. at 5000 ft. (450 h.p. at 2300 r.p.m. for takeoff); length overall 30'4"; height overall (on wheels) 8'7"; wing span 48'0"; wing chord 62.5"; total wing area 250 sq. ft.; wt. empty 2810 (3000) lbs.; useful load 2010 (2100) lbs.; payload with 95 gal. fuel 1225 (1315) lbs.; gross wt. 4820 (5100) lbs.;

"Beaver" in Alaska; a shabby veteran of bush-flying.

The high-performance "Turbo-Beaver" had turbine engine.

multiple figures show wts. as allowed with landing gear modifications; max. speed 160 at 5000 ft.; redlined at 195; cruising speed (.60 power) 130 at 5000 ft.; landing speed (with flaps) 45-50; stall speed (no flaps) 55-60; takeoff run (no wind) 560 ft.; takeoff run (over 50' barrier) 1100 ft.; takeoff run on skis 680 ft. on average; takeoff run on floats 890 ft. on average; climb 1020 (835) ft. first min. at SL; landing roll (with flaps) 450 ft.; landing run (over 50' barrier) 1000 ft.; ser. ceiling 18,000 ft.; gas cap. 95 (US) gal.; optional gas cap. 138 (US) gal.; oil cap. 6.25 gal.; cruising range (.60 power at 5088 ft.) using 18-20 gal. per hour was up to 600 miles; prices started at $22,000 in 1948 and rose progressively to $28,000 and then to $49,500 in 1963; 1967 models sold for $42,000 less engine! With Edo 4820 twin-float gear the gross wt. was limited to 5090 lbs. The U-6A was also eligible as DHC-2 when modified to conform.

The fuselage framework was a robust semi-monocoque structure of various dural (alum. alloy) members and was covered with stressed-skin alum. alloy "clad" sheet. The floor was reinforced for heavy loads and the interior was serviceable, but nothing fancy. The cabin actually had room for 8 seats (2 front, 3 center, 3 rear) and all were quickly removable to allow 125 cu. ft. of cargo space for the loading of rather bulky items. For main cabin entry there was a large door and step each side, and for cockpit entry there was a smaller door and 2 small steps each side. The fair-sized interior was certainly not appointed like a limousine, but it was especially adequate for hard work; an 8 cu. ft. baggage compt. was to the rear of the cabin area. Three fuel tanks of 35-35-25 gal. (US) capacity were mounted in the belly of the fuselage for easy servicing; oil tank filler-neck was in the cockpit. The slender semi-cantilever high aspect-ratio wing was built up of heavy dural spar beams, dural wing ribs, and was covered with stressed-skin alum alloy "clad" sheet; the slotted wing flaps and ailerons were also covered in dural metal sheet. The wing flaps were operated by hydraulic hand-pump to 48 degs.; no flaps above 105 m.p.h. Ailerons were interconnected to wing flaps for "droop" to provide more low-speed lifting area. The cantilever landing gear was a particularly robust structure using rubber-donut and oil-spring shock struts, and 7.50x10 wheels with hydraulic brakes; a parking brake was also provided. Increased gross wt. to 5100 lbs. was later allowed with 8.50x10 wheels and tires; skiplane allowed gross wt. to 4820 lbs. The 5.00x4 swiveling tail wheel was also steerable. Various types of skis were used in the winter off snow, and Edo 58-4820 twin-float gear was usually used in the summer; amphibious floats were also used in some areas. The all-metal cantilever tail group was also a semi-monocoque structure using rudder & elevators with aerodynamic balance; the elevators had adj. trim tab, while rudder & ailerons had fixed tab adj. on the ground only. A Hamilton-Standard constant-speed prop, engine starter, generator, 24V battery with full electrical system, oil cooler, fuel pumps, carburetor heater & air-filter, normal set

"Beaver" dons ski-gear for the winter.

of engine & flight instruments, compass, airspeed ind., cabin heater, cabin vents, navig. lights, landing lights, & tie-down rings were std. equipment. Special interiors, long-range fuel tanks, skis, pontoons, oversized wheels, radio gear, crop-dusting equipment, canoe-carrying eqpt., lumber-carrying eqpt., extra instruments, fire extinguisher were among the options available. Seaplane was required to have aux. fin; floats added 307 lbs. to the normal empty wt. Floats for the "Beaver" cost over $20,000.

It was noted that 363 of the "Beaver" were still on the U.S. register as of 3-31-80; 154 were listed as the DHC-2, 11 were the U-6, 3 as L-20, 179 as the U-6A, 5 as the L-20A, and 11 as the DHC-2 Mk. 3. Of these above, 87 of the "Beaver" were registered in Alaska where they operate in season on wheels, skis, or floats. About 100 of the total were registered with various government depts. & agencies; the ones in the U.S.A. were pretty well scattered around the country, with perhaps a larger portion in the west.

ATC #807
(1948)
DE HAVILAND "DOVE," DH-104

The lovely DeHaviland "Dove" with DH "Gipsy" engines.

Seeing the need of a logical replacement for the hard-working "Dragon," "Rapide" and "Dragonfly" biplanes of prewar fame, DeHaviland Aircraft began designing the "Dove" as their first post-war product. Losing no time in formulating its final design, the prototype "Dove" made its first flight only 6 weeks after the war ended on 9-25-45 with Geoffrey Pike at the controls. There was the usual development testing which dictated a modification here and there, and in June of 1946 it received its British certificate of airworthiness (C of A). Being fitted with (DH) DeHaviland engines, and DeHaviland propellers, it was the first "DH" airplane to be built almost entirely by DeHaviland Aircraft. Because orders were at first slow in coming from domestic British lines, DeHaviland made a concerted drive for export business, and it began paying off. Large scale production was begun in 1947 and subsequently, deliveries were soon being made to the Sudan, to West Africa, So. Africa, Kenya, Nigeria, So. Rhodesia, Malta, Iraq, the Persian Gulf, Kuwait, India, Burma, and the French colonies. By June of 1948 some 100 of the "Dove" were already in service and over 200 more were on order. After establishing an early worldwide reputation for profitable short-haul service, a gaggle of them even made their way to Canada and America; they flew the route via Prestwick and Gander. In the summer of 1957 the TAG Airlines (Detroit to Cleveland) started service in the Great Lakes area with two 9 passenger "Dove"; the service became so popular and profitable that before long they had 11 of the "Dove" on the flight-line operating approx. 200 hours per month. Well-suited to this kind of treatment, a "Dove" could be seen taking off every 15 minutes! TAG Airlines lauded the dependability and flexibility of the DH "Dove," a salutation that was echoed almost all around the world. As the sweet-running "Gipsy" inverted inline engines began to show their age they were usually replaced here with American engines of the 6 cyl. opposed type, engines such as the Continental or Lycoming. In general, this engine switch added horsepower and improved the performance. In 1963 the Riley Aeronautics group announced it would convert the (DH-104) "Dove" by installing 400 h.p. Lycoming engines, or 562 h.p. turbo-prop (turbine) engines; this must have made the stately "Dove" into a real charger! It has been said that more than 500 of the "Dove" were built in all, and chances are good that most of these, modified and converted, are still flying here and there around the world.

The graceful DeHaviland (DH-104) "Dove" was a light twin-engined, all-metal cabin monoplane with variable seating arrangements to suit a variety of services. The basic "Series 1" had seating for 10 to 13 passengers and a crew of two, while the Series 2 with seating for only 6 to 8 passengers and a crew of 2 was the so-called "Executive" model. Incidentally, the Series 1A and 2A were specially fitted for export to America. The lovely "Dove" was a distinctive airplane that looked rather shy and lady-like, but its looks actually belied all the muscle and ability that posed within her frame. The twin-engined "Dove" was not particularly a bush-type airplane in the true sense, but it actually was designed to use the smaller airfields in undeveloped areas for short-haul routes on a frequent schedule. The two smaller engines provided a good measure of safety, and also an ample power reserve; it had better-than-average performance with "one engine out," and could operate efficiently on less than 2 cents per passenger-mile. As powered with two 6 cyl. air-cooled inline DeHaviland "Gipsy Queen" 70-4 engines rated 330-340 h.p. the "Dove" was quite a lively ship and its overall performance was quite impressive; whether difficult or easy, it took on all chores with a measured stride of confidence. Flight characteristics were always described as pleasant, and her overall nature was lady-like and gentle, but she could become a "tomboy" real quick, they say, if the occasion called for it. In other words, she responded well to varied situations. Pilots had nothing but a good word for the "Dove," and many will say that the hardest part in flying a "Dove" regularly

TAG Airlines found "Dove" ideal for their service.

British-marked "Dove" shows its grace in flight.

"Dove" as someone's air-yacht, named "Cloud 9.

was getting in and out of that tiny little cockpit. Being of rugged structure too, the "Dove" wore quite well over the years, wearing out many engines, and quite a few are still out there flying. The type certificate for the "Dove" in America seems to have been issued sometime in 1948 and altogether some 500 or more examples (of the various Series) were mfgd. by the DeHaviland Aircraft Co., Ltd. on Hatfield Aerodrome in Herts, England. Approval was later awarded to Series 5A and 6A which were basically same as the 1A and 2A, except with engines of 355-380 h.p. and gross weight raised to 8800 lbs. The Series 1A and 2A were also eligible for conversion to 5BA and 6BA if operators wanted to use the more powerful engines, but didn't necessarily need the extra allowance for gross weight.

Listed below are specification and performance data for the DeHaviland (DH-104) "Dove" as powered with two 6 cyl. DH "Gipsy Queen" 70-4 engines rated 330 h.p. at 2600 r.p.m. at 5000 ft. (340 h.p. at 2800 r.p.m. for takeoff); length overall 39'4"; height overall (at rudder) 13'0"; wing span 57'0"; wing chord tapered in planform & section; wt. empty 5625 lbs.; useful load 2875 lbs.; payload with 100 gal. fuel & 2 crew 1835 lbs. (10 pass. & 135 lbs. bag.); gross wt. 8500 lbs.; max. speed 220 at SL; cruising speed (.60 power) 155 at 5000 ft.; landing speed (with flaps) 68; stall speed (no flaps) 75; climb 750 ft. first min. at SL; ser. ceiling 18,000 ft.; gas cap. 142 (US) gal.; oil cap. max. 20 (US) gal.; cruising range (.60 power at 5000 ft.) using 27 gal. per hour was 450 miles with 100 gal. fuel; price $71,500 in U.S.A., & 17,500 pounds sterling (British) with std. equipment. Series 5A and 6A with engines of higher horsepower was allowed 8800 lbs. gross wt.; performance was about equal.

The fuselage framework was an all-metal semi-monocoque structure of dural (alum. alloy) members covered with "clad" metal sheet. A cabin entry door was on the left side behind the wing, and single seats inside were arranged the length of the cabin either side of a center aisle. Visibility was excellent and every one had a window seat. The pilots also had good visibility in the elevated dome-like cockpit, a carry-over from earlier DeHaviland designs. The versatile interior could be arranged at just about any density to suit a particular service, and could be made as utilitarian or deluxe as desired. With all cabin seats removed there was 89 cu. ft. of cargo capacity with suitable tie-downs. The graceful cantilever wing was built up with dural spar beams, stamped dural wing ribs, and covered with stressed-skin (alum. alloy) "Clad" metal sheet; ailerons and wing flaps were covered in fabric. Fuel tanks were mounted in the wing's center section, two each side of the fuselage. The retractable tricycle landing gear with rubber-donut shock struts used 27x8 main wheels with air-operated brakes; landing gear operations (up & down) was also pneumatic. The 7x6 nosewheel was steerable. The cantilever tail group of distinctive DeHaviland outline was an all-metal structure with fixed surfaces covered in "clad" metal sheet, and movable surfaces were covered with fabric. All movable surfaces were aerodynamically balanced and had adj. tabs for three-way trim. The DeHaviland (Hydromatic) 3-bladed props were variable-pitch, full-feathering, and were reversible for braking upon landing. A 24V electrical system, battery, generator, engine starters, oil coolers, fuel pumps, air pumps, a normal set of engine & flight instruments, cabin lites, cockpit lights, navig. lights, landing lights, compass, airspeed ind., parking brake, cabin appointments, cabin seating, and cabin hardware were std. equipment. Custom interiors, lavatory, radio gear & airline equipment were optional.

It was noted that 51 of the DeHaviland "Dove" were still on the U.S. register as of 3-31-80. Four were the model 1A, 3 were the 2A, and the balance of the total were divided about equally for the models 5A and 6A. One "Dove" was registered in Hawaii.

A "Dove" in Canada—popular there too.

ATC #808
(9-23-48)
CURTISS-WRIGHT "COMMANDO," C-46F

A converted C-46F on the line for Meteor Air Transport.

The huge double-decked C-46 "Commando" transport of World War 2 fame was built in almost 3200 units and delivered to the USAAF from 4 different plants in more than 10 different versions. Shortly after the war a good number of these were declared surplus to military needs, and more than 600 were put up for sale. Because the airplanes were really a good bargain their use was ideal for the many non-sked (non-scheduled) airlines that were springing up all over the country. It seemed, then too, a lucrative endeavor to buy up these old warriors, at a mere fraction of their original cost, and convert them to civil use. Most were fitted to haul perishable and packaged air-freight, altho' there were a few passenger-carrying conversions also. In time, six different type certificates were issued to various remodelers of which Slick Airways was the first; Curtiss-Wright held the certificate to remodel dolphin-nosed (or broken-nosed) C-46E, and USAIR held a certificate to remodel the C-46A, D, F. This particular certificate (#808) was issued to Skyways International Trading & Transport Co. of Miami, Fla. to convert the popular C-46F for civil use. There were only 234 units of this C-46F version built altogether; Curtiss-Wright had initial orders for 400 units, but 166 were cancelled on VJ day. The amount left to be sold off as surplus was probably less than 200 airplanes. In a quick run-down thru' pertinent material there were no available figures for the amount of airplanes that Skyways International had converted; both Slick Airways and USAIR had converted the C-46F model also, so the amount converted or remodeled by any one of these remodelers could not have been very many. The USAAF also leased about 100 of the various C-46 versions to 30 civil carriers from 1947 thru' 1949; among these lines were Pan Am, Delta, Eastern, and National. All of the airplanes were returned to the government by 1953. As these converted and leased C-46 transports were put into the mainstream of civil air transport, they generated a significant amount of history into post-war aviation. Air transportation had quite a time of it as it tried to rebuild itself after the second World War.

The famous Curtiss-Wright "Commando" of the Army-type C-46F series was a large all-metal, low-winged, twin engined transport monoplane; it was fashioned with a cavernous interior that was highly suitable for the hauling of heavy cargo, air-express shipments, and other varied tonnage. As a military-type transport (no civil versions were ever delivered from the factory) it fell heir to just about all the back breaking chores that were around during war-time, and in this capacity it did a pretty fair job, even under trying circumstances. Pilots have cussed and discussed this airplane for many years, and historians are sure to do likewise. As powered with 2 P & W "Double-Wasp" R-2800 engines of some 1700-2000 h.p. each the loaded C-46 was certainly not overpowered for what she was asked to do in war time, but the civil conversions were treated a little more kindly. Even then operation of this big airplane had to be "heads up" for best results. For a big, heavy, rugged airplane that was very often mistreated, she flew well under conditions that were not always ideal; when everything was going good, and the crew was on top of things, the ol' C-46 was rather hard to beat. Because of its higher cruising speeds, and low ton-per-mile operating costs, the C-46 was generally preferred over the C-47 (DC-3) type for hauling cargo. With a sturdy basic structure designed for hard service in primitive areas it is not surprising that these airplanes were still flying for decades even after they had been declared war-surplus. A few of these (C-46) hulks are still sitting around here and there, slowly decaying in the elements, but it wouldn't take all that much fixin' to make them fly again. The big "Commando" made a name for itself during war-time as the "Troop Ship of the Sky," and in carrying badly-needed war supplies "Over The Hump" to China; as a "civilized" version she did almost as well in her new role. The stories that revolve around this airplane are hair-raising, unbelievable, and sometimes comical—it would make interesting reading. Following phase-out of their general use in this country the C-46 meandered up into Canada and down into Latin America; some of these might be flying yet. The type certificate for conversion of the C-46F type airplane (in this instance) was issued 9-23-48 to Skyways International Trading & Transport Corp. of Miami, Fla. It is worth noting that L.D. Smith held Type Certificate 2A5 for C-46 and CW-20T conversions and Riddle Air Lines held TC 3A2 for converting C-46A, D, F, and R models.

Listed below are specifications and performance data for the C-46F (conversion) as powered with two 18 cyl. "Double-Wasp" R-2800-75, -51, -53 engines rated 1700 h.p. at 2550 r.p.m. at 5500 ft. (2000 h.p. at 2700 r.p.m. for takeoff); length overall 76'4"; height overall (at rudder) 21'9"; wing span 108'0"; wing chord at root 198" tapering to 5'6" at tip; total wing area 1366 sq. ft.; airfoil NACA-23017 at root tapering to NACA-4410.5 at tip; wt. empty 27,000 lbs.; useful load 17,000 lbs.; payload with 1406 gal. fuel, 80 gal. oil, & 2 crew 8580 lbs.; normal gross wt. 44,000 lbs.; provisional gross wt. to 45,000 lbs.; max. speed 240 at 10,000 ft.; redlined at 270; cruising speed (.60 power) 215 at 10,000 ft.; landing speed (with flaps) 74; stall speed (no flaps) 85; takeoff run 2200 ft.; climb 1500 ft. first min. at SL; landing run 2150 ft. to stop; ser. ceiling 25,000 ft.; gas cap. 1406 gal.; oil cap. 79.6 gal.; cruising range (.60 power at 10,000 ft.) us-

A C-46F all decked-out for U.S. Airlines.

This view shows many details of the C-46F type.

ing 180 gal. per hour was 2000 miles; fuel was often traded off for more payload and less range; single-engine ceiling was about 9000 ft. It was noted that the C-46F cost the government $233,370 when new, while they were sold off as surplus for from $5000 to $25,000!!

The construction details and general arrangement of all the C-46 type was more or less the same from model to model except door location, size of door, interior arrangements, and a number of slight differences that were not always easily visible. On the C-46F the cargo door on the left side was 6 ft. high x 8 ft. wide with sometimes a smaller insert to avoid opening the whole large door; the floor was reinforced with a max. floor loading to 185 lbs. per sq. ft. All cargo compts. or bays were marked for allowable weight limits. The passenger-carrying version was limited to a max. allowance of 62 passengers when modified to prescribed directives; 3 emergency exit doors beside the main entry door were required. The fuel load was divided into 6 tanks, 3 in each wing; 236 gal. in forward tank, 292 gal. center, and 175 gal. in rear tank. No fuel or oil tanks were allowed in the fuselage. Wing flap deflection was limited to 35 deg.; no flaps above 135 m.p.h. and no landing gear extension above 150 m.p.h. The landing gear of 311" tread used "Aerol" shock struts with 19.00x23 wheels fitted with hydraulic brakes; tail wheel was 10.00x7. The required crew for this conversion was 2 pilots and a flight engineer. The cabin

A C-46F in Canadian service.

heater was a gasoline-burning device; refrigeration and air-conditioning was also available. These conversions were usually equipped with four-bladed Curtiss Electric props, but certain Hamilton-Standard props were optional. A de-icing system, anti-icing system, 3-way trimming control, a 24V electrical system, 24V batteries, a fire extinguishing system, bonding & shielding, & a full set of airline equipment was usually standard. Many options were available to customer specs., if not contrary to government directives. All ser. nos. of Army-type C-46F airplanes were eligible for this conversion approval held by Skyways International; nameplate bearing conversion date must be installed adjacent to factory nameplate.

ATC #809
(CANCELLED)

ATC #810
(CANCELLED)

ATC #811
(CANCELLED)

It is quite unusual that these three "type certificates" (#809-810-811) cannot be accounted for. After several months of research by myself, and many other noted historians, it has been determined with the help of the FAA (Federal Aviation Authority) that these 3 certificates were more than likely awarded, but never used. Then too, it cannot be proven either that these "approvals" were ever issued, otherwise there would certainly have to be some record, somewhere. In years past, it was not unusual to reserve a "type certificate number" for an airplane that was undergoing tests for certification at the time. If the airplane, for one reason or another, was not subsequently certificated (or approved) the ATC number was sometimes cancelled and in any case, was not used again for another airplane.

The first time this had happened in more than 20 years was for ATC #798 as explained in U. S. CIVIL AIRCRAFT, Vol. 8. The likelihood that this would happen again was thought to be very remote, but it did happen again, and for three "approvals" in a row! Because of this the government agency then discontinued the practice of awarding numbers prematurely, and no "number" was then allocated until the certification tests were completed. Because seven of the numbers between #801 thru' #817 were awarded to imported aircraft it was assumed that #809-810-811 might have been awarded to foreign-built airplanes, but the FAA stated that this need not be necessarily so. Because this mix-up, or failure to record properly, came at a time when "type certificate" allocation was being changed over from a "numerical" to a "regional system." the records may have been lost in the confusion. Somewhere in a mound of papers, or a dusty file, lies the answer perhaps, and the chore to find this may be likened to "hunting for a needle in a haystack"!

ATC #812
(9-3-48)
BOEING "STRATOCRUISER," 377

Boeing "Stratocruiser" featured most luxurious air-travel ever devised.

The sprawling Pan American World Airways system, already responsible for the design and development of many fine airliners, had made known its requirements to several manufacturers in 1941 for a deluxe long-range postwar transport; it would have to have a payload of at least 9 ton, with a cruising range of some 5000 miles, and show top speeds close to 375 m.p.h. With the way already paved somewhat by the prewar "Stratoclipper" (Model 307) Boeing made its bid with a new design in 1942. This design eventually made its first appearance as the military "Strato-freighter" (XC-97) which was largely based on the famous B-29 bomber, in the same way as the Model 307 was based on the war-time B-17 bomber. The new design used wings, engines, and tail group of the B-29 to which was fitted a huge figure-8 fuselage that was pressurized, air-conditioned, cavernous, and lavishly comfortable. The prototype XC-97 (Model 367) was up on its maiden flight in Nov. of 1944 and its success prompted Boeing to announce its civil "Stratocruiser" (Model 377) that was being developed for airline use; it was already in mock-up by Nov. of 1945. The commercial "Stratocruiser" was largely based on the YC-97 design having a double-decked cabin that would carry 50 to 80 passengers in luxurious comfort; it also provided a downstairs lounge where passengers could enjoy a cigarette and a cocktail or two. With forethought of the type of service they would render, Pan Am suggested the use of the newly-developed 3500 h.p. "Wasp Major" (TSB3G) R-4360 engines then being tested by Pratt & Whitney, so the "Strato" then had powerplants and nacelles similar to those used on the new B-50 bomber. Maiden flight of the first civil "Stratocruiser" was on 7-8-47, and the second "Strato" first flew in Aug. of 1947. Production of this airliner, for which Boeing already had 55 orders, followed the 10th airplane of the YC-97 contract, so the "Stratocruiser" (377) and the B-50 bomber were being built side-by-side on the assembly line.

First put into service by Pan Am on its San Francisco to Honolulu run in April of 1949 the "377" soon proved that it could fly faster, farther, and higher, with more luxury aboard than any airliner built previously. As the largest land-based transport airplane in the world it offered room, convenience, and smile-provoking luxury; truly first-class travel as never before experienced by passengers or the crew. For this it was remembered long after it had been replaced on major airlines by newer and faster equipment. Pan Am ordered 20 of the "Stratocruiser," 10 were delivered to Northwest Orient Airlines, and some 25 were ordered by 4 other airlines; among these were British Overseas Airways, American Overseas Airways, and of course, United Air Lines. TWA ordered also, but soon cancelled its order.

Put into service amid much fanfare, with christenings by lovely ladies and famous personalities, and with names borrowed from great sailing ships, the "Stratocruiser" was soon making Pacific and Atlantic Ocean crossings on schedule. The service was very popular and people vied with each other to get a booking; this was an adventure worth trying and bragging about. The huge "Strato" continued to give popular and illustrious service for the best part of a decade, and only then was made obsolete by the coming of the jet-liners of that period. But, their service certainly did not end there because the "377" were bought up by other airlines, mostly foreign carriers, and they continued to haul passengers and cargo to all corners of the world. Meanwhile, its military sister-ship the C-97 series, were blazing trails in the new art of in-the-air refueling, the transport of fully-equipped troops to far-flung places, and hauling such sundry cargo as coal during the now-famous Berlin Air-Lift. Some of the "Stratos" were later modified for unusual or special chores; the most outlandish modifications were for Aero Spacelines and popularly known as the "Pregnant Guppy," the "Super Guppy," and the "Mini-Guppy." These specialized airplanes carried (Saturn) space-vehicle components from west-coast manufacturers to test-sites in Florida.

The huge Boeing "Stratocruiser" Model 377 was a four-engined long-range airliner that seated anywhere from 55 to 80 passengers plus a crew of 7 to 9 people, in a comfort not often experienced by air-travelers before this. Its interior space allotments, for both passengers and the crew, was so generous that it promoted calm and friendliness into everyone aboard. The bulky "Stratocruiser" was certainly no svelte-looking beauty on the outside by any means, but on the inside she surely was a people-pleaser. In the first decade of her service she blazed new trails, set various records, and was the favorite of everyone, however, she was rather costly to operate and quite expensive to maintain properly; this had finally caused its obsolescence on the major airlines. All 55 of the "377" ordered by the airlines were delivered between Feb. of 1949 and March of 1950 at a price, more or less, of 1½

"Stratocruiser" being honored in Hawaii after its inaugural flight.

"Strato" high above Atlantic on way from Europe.

million dollars per airplane! Altho' the 377 delivered to 6 different airlines originally were structurally the same they differed considerably in interior appointments; each airline had its own ideas for the arrangement and styling of the interior decor. Each brand of particular styling was then identified in the model designation by a "dash number." When finally retired from major-airline service the "377" was generally modified to serve other airlines for plush charter-flights, as a high-density transport seating up to 117 paying passengers, or as an all-cargo carrier. Perhaps the strangest modifications ever were the famous "Guppy" series as used by Aero Spacelines.

As powered with four of the Pratt & Whitney (TSB6G) R-4360 "Corn-Cob" engines of some 3500 h.p. each, engines made up of 28 cyls. arranged in 4 rows of 7, the "Stratocruiser" was a good performer in spite of its great bulk and heft. It took a certain amount of understanding to master its operation properly, but most of the pilots liked it. The passengers were thoroughly capitavated by the pleasant atmosphere and the enjoyable ride; the crew were happy for the conveniences that made their jobs more pleasant. If everything was going well the flying and operating of the "Strato" was a "piece of cake," but things could get pretty hairy quickly if all was not well. The huge "Cruiser" was a proud lady that had spanned oceans for a decade or more, and its conspicuous arrival was always cause for gatherings, but its majesty was finally overshadowed by the coming of "the jet age" which Boeing helped introduce. The type certificate for the Boeing "Stratocruiser" Model 377 was issued 9-3-48 and some 56 examples (including the prototype) were mfgd. by the Boeing Airplane Co. of Seattle, Wash.; all production of the Model 377 was at the Seattle facility. Clair L. Egtvedt was chrmn. of brd.; Wm. M. Allen was pres.; Wellwood E. Beal was V.P. of engrg. & sales; Fred Collins was sales mgr.; Edw. C. Wells was chf. engr.; & N. D. Showalter was chf. of flight-test. John B. Fornasero later became chf. of flight-test.

Listed below are specifications and performance data for the "Stratocruiser" Model 377 as

"Stratocruiser" impressed everyone with its size.

Some intimate details of the "377."

powered with four 28 cyl. "Wasp Major" R-4360 (TSB3G or TSB6G) engines rated 2800 h.p. at 2550 r.p.m. at 5500 ft. (3500 h.p. at 2700 r.p.m. with water injection for takeoff); length overall 110'4"; height overall (at rudder) 38'3"; wing span 141'3"; wing chord tapered in planform & section; total wing area 1769 sq. ft.; airfoil "Boeing 117"; wt. empty 83,500 lbs.; useful load 59,000 lbs.; max. payload 30,000 lbs.; payload with 7790 gal. fuel, 194 gal. oil, & 7 crew was 9400 lbs. (45-50 pass. & baggage); normal gross wt. 142,500 lbs.; provisional gross wt. later increased to 145,800 & 147,000 lbs.; max. dive speed 390; max. speed (level flight) 340; cruising speed (.66 power) 315 at 25,000 ft.; landing speed (with flaps) 97; stall speed (no flaps) 115; max. speed with flaps extended 180; climb 1000 ft. first min. at SL; ser. ceiling 25,000 ft.; max. gas cap. 7790 gal.; oil cap. 194 gal.; cruising range (.66 power at 25,000 ft.) using 467 gal. per hour was approx. 4600 miles; price from $1,250,000 to $1,500,000 with std. equipment.

The huge pressurized, figure-8 fuselage was a semi-monocoque structure largely of 75ST alum. alloy fashioned in a technique that was the latest in the state of the art; the completed structure was then covered with stressed-skin "Alclad" metal sheet. Distinctive features of the Model 377 were the double-decked fuselage of some 10.5 ft. in diameter, and its many-windowed cockpit which featured some 2800 sq. ins. of glass area; the flight-crew which normally numbered 5 were all together up on the spacious flight-deck. The upper deck of the cabin area was arranged to suit various airlines, but was usually arranged with a private state-room, a washroom for both men and women, a main cabin area that seated from 45 to 80 passengers, an optional "sleeper" section that had 28 berths and 5 recliner chairs, and the fully-equipped galley was generally in the rearmost section; United Air Lines had the galley in the middle to divide the airplane into 2 different rooms. A spiral staircase led to the cocktail lounge in the lower deck; the lower deck also housed the baggage, misc. cargo, and the life-rafts. Boeing offered 11 different interior configurations. Entrance to the forward lower cargo deck also provided entry to the cockpit; passenger entry was amidship on the left side. Luxury was the keynote in transporting Mr. & Mrs. Public by air; the cabin contained all known elements of comfort including a man-made climate. Because the huge 71 ton liner was most efficient at 15,000 ft. or higher, the cabin was pressurized to maintain a sea-level "altitude" up to 15,000 ft. at a comfortable 70 degs. At 25,000 ft. the cabin altitude was at 6000 ft. The huge engine nacelles were a complex package that included an auto. fire extinguishing system. Four-bladed Curtiss Electric, or Hamilton-Standard props were used; the Ham-Std. props were more efficient, but more prone to malfunction under severe use. The retractable tricycle landing gear of 342 in. tread used 56x19 dual Goodrich wheels with hydraulic brakes; the dual 36 in. nosewheel was steerable. The huge tail group reached some 38'3" into the air; to clear a hangar roof this assembly could be folded down to 26'7". The rudder, elevators, and ailerons had adj. tabs for 3-way trim. Huge variable-area "Fowler Flaps" improved the takeoff and slowed the landings. The equipment included was very extensive and complied to the latest airline specifications; this included radar, radio gear, de-icing & anti-icing equipment, thermal de-icing for the wings & tail group, heated windows at the pilot station, 24V electrical system, landing lights, & auto. pilot. No "377" are operational in this country any more; a retired C-97 rests in Dodgeville, Wis. where it has been converted to a deluxe coffee-shop as part of the "Don Q" motel complex.

Lower-deck cocktail lounge offered relaxation in flight.

ATC #813
(8-15-55)
TRECKER (PIAGGO)
"ROYAL GULL," P.136-L

Royal "Gull" was a distinctive and very efficient airplane.

Typical of the artistic flair shown in most all their doings, the Italian designers, both past and present, have created some of the most beautiful airplanes in the world. They certainly had and have a way with lines and form that was and is classic. A very good example of this was the beautiful Piaggo "Gull." The "Gull" (P.136) by Piaggo, who had been building airplanes since back in 1915, would be in the same general class you might say as the Grumman "Wigeon," but there the comparison ends. The lovely "Piaggo" with its gracefully "gulled" wing, its highly-streamlined "pusher engine" installation, and its smooth-flowing hull was crafted to appeal; but, it was also to be one of the most efficient amphibious airplanes in the world. Labeled as the P.136 the prototype Piaggo airplane first flew in Aug. of 1948, and the first 20 or so went to the Italian Air Force; there to do the search-rescue and patrol work that only a good amphibian can do. A few civil-commercial versions were also sold, and one of these went to Aristotle Onassis (multi-millionaire) who lashed it aboard his famous yacht "Christina." Forseeing a possible market for the "Gull" in the U.S.A., a land of millionaires they assumed, Piaggo negotiated with several manufacturers, but they had to knock on quite a few doors to find an interested party. The nod finally went to the Royal Aircraft Corp. of Milwaukee, Wis.; Royal Aircraft was a subsidiary of Kearney & Trecker the well-known manufacturer of "Milwaukee" milling-machines and other fine machine tools.

The first 3 airplanes came completely assembled from Genoa, Italy by boat to New York City, and Carl G. Koeling, chief pilot for Kearney & Trecker, flew them to Milwaukee; 29 more later came in crates to be assembled here. Under the agreement the major airframe components were built by Piaggo then brought over by boat directly to the port of Milwaukee, and assembled by Royal Aircraft for sale and delivery. Of course, much of the airplane took on an American flavor as it went together because it was equipped with engines, props, many operating components, and accessories manufactured in this country. Deliveries began in 1955, but no one was beating a path to the company's door; there were orders, of course, mostly from business-houses, but they were few and far between. There was nothing else in its class at this time, but the "Gull" was definitely a rich-man's airplane, so its sales potential was rather limited. A few were sold in Canada. For whatever reason the Royal Aircraft Corp. was then reorganized into the Trecker Aircraft Corp. and an improved model marketed, but that didn't create no big splash either. Out of the 75 or so that were built in total, some 32 of the "Gull" were built by Royal and Trecker; more than one-half of these are still flying. Trecker Aircraft discontinued the aircraft-assembly operation in 1960, and dissolved entirely in 1964.

The softly-rounded, sexy-looking Royal (Trecker) "Gull" P.136 was a twin-engined high-winged cabin monoplane of the "amphibian" type. There was ample room for five people, and the airplane could operate equally as well off land or water. It was such a beautiful and striking airplane that famous business-firms and even individuals, bought it mostly for the show-off value and the prestige. There were many innovations on this airplane and the unusual "gulled wing" was not just a gimmick—it was designed to keep the engine high out of the water-spray. The "pusher engines" also put the noise and vibration away from the cabin where the people could find peace and quiet along with their comfort. Being mainly a "boat," the "Gull" had excellent water characteristics; and, its operation on land was not as wobbly as most airplanes of this type. Everything on this airplane was first-class, and it had a price-tag to match. As powered with two 6 cyl. geared Lycoming GO-480-B1B engines of 260-270 h.p. each the "Gull" (P.136-L1) was a spritely machine that was astonishingly deft for an airplane of this type. Take-offs and landings required but very little space (both on land and water), so the "Gull" could actually visit the most interesting places. Altho' it handled pretty much like a big ship, its flight characteristics were very pleasant and its overall nature was about as lovely as its good looks. There was plenty of room for everything and everybody because Piaggo did not skimp on room, nor did they skimp on material or the construction. Because they were tough and reliable more than half of those "Gull" built are still flying. And, the market for the ageless "Gull" is better now than it was 25 years ago! The type certificate for the model P.136-L was issued late in 1954, with amendment later for the P.136-L1; approval for the higher-powered P.136-L2 was

The lovely "Gull" owned and flown by noted writer "Ernie" Gann.

A striking and amusing view of the Piaggo "Gull."

issued on 3-7-57. Some 32 examples of the "Gull" were mfgd. by the Royal Aircraft Corp. and the Trecker Aircraft Corp. on General Mitchell Field, both subsidiaries of the Kearney & Trecker Corp. of Milwaukee, Wis. Carl G. Koeling was the chief pilot who tested the airplanes as they came off the line.

Listed below are specifications and performance data for the Royal "Gull" model P.136-L1 as powered with two 6 cyl. geared Lycoming G0-480-B1B or -B1D engines rated 260 h.p. at 3000 r.p.m. (270 h.p. at 3400 r.p.m. for takeoff); length overall 35'6"; height overall (at rudder) 11'6"; wing span 44'5; wing chord tapered in planform & section; total wing area 270 sq. ft.; airfoil NACA-230 Mod.; wt. empty 4400 lbs.; useful load 1600 lbs.; payload with 100 gal. fuel 784 lbs. (4 pass at 170 lbs. each & 105 lbs. bag.); gross wt. 6000 lbs.; max. speed 183 at SL; redlined at 222; cruising speed (.65 power) 164 at 8000 ft.; min. controllable speed 90; landing speed (with flaps) 68; stall speed (no flaps) 75; takeoff run (no wind off water) 18 secs.; takeoff run (on land) 965 ft.; landing run (with flaps) 760 ft.; climb 1180 ft. first min. at SL; ser. ceiling 18,500 ft.; normal gas fill 100 gal.; max. gas cap. 190 gal.; total oil cap. 6 gal.; normal cruising range with 100 gal. fuel (.65 power at 8000 ft.) using 30 gal. per hour was 480 miles; max. cruising range with 190 gal. fuel was 960 miles; the factory advised "call or write for price." Performances of the earlier P.136-L was only slightly inferior; the later P.136-L2 with 2 geared & supercharged 6 cyl. SGO-480-A1A6 engines rated 320-340 h.p. did show a greater performance with increased gross wt. to 6393 lbs. (sea) and 6614 lbs. (land). Service ceiling for the P.136-L1 with one engine dead was 4100 ft.; the figure for earlier & later models would be similar.

The hefty all-metal hull was of semi-monocoque construction with frames, the bulkheads, keelsons, and stringers covered with stressed-skin alum. alloy "clad" sheet. The graceful hull was a novel two-step design that distributed water-horne weight over a larger area, and its length was divided into 7 watertight compts. The cabin up forward for 5 people had ample window area, with a large baggage compt. just behind the rear seat; a large baggage-cargo hold aft had an outside access door on the left side. Large entry doors were on either side up front, and right panel of the windshield could be swung open to better manage mooring and docking. The soundproofed interior was done up comfortably in serviceable leather decor; adj. seats up front had folding backs, and the rear bench-type seat was wide enough to seat 3 across. The "gulled" cantilever wing was an all-metal semi-monocoque structure covered with stressed-skin alum. alloy "clad" sheet; the "backwards" engine nacelles were built right into the wing. A 95 gal. fuel tank was mounted each side in the wing root; fixed tip-floats were attached to the wing by a single cantilever strut at a point out to prevent "heeling" into the water. Landing lights were mounted in the wing's leading edge on each side. The hydraulic retractable landing gear folded up into a well alongside the hull, and the tail wheel folded up out of the way. Wing flaps or landing gear not to be lowered above 126 m.p.h.; flap deflection was to 45 deg. The tail group was

A "Gull" poised on water's edge.

Trecker "Gull" featured larger engines and a dorsal fin.

also an all-metal cantilever structure; the P.136-L2 was fitted with a large dorsal fin because of the higher-power engines. All control surfaces were aerodynamically balanced; the elevators and rudder had adj. trimming tabs. Three-bladed constant-speed full-feathering Hartzell metal props, normal set of engine & flight instruments, compass, airspeed ind., vacuum pumps, fuel pumps, hydraulic pumps, 24V battery, engine starters, 2 generators with separate electrical systems that could still operate everything with one engine dead, radio gear, wheels, tires & brakes were all American-made equipment. Other equipment included navig. lights, landing lites, clock, dome lights, cabin heater, cabin vents, armrests, window curtains, floor rug, dual control wheels, dual brake pedals, parking brake, map-pockets, anchor & mooring gear, ash trays & tie-down rings. There were several unspecified options available at extra cost. Std. exterior color was Alaskan White with two-tone Green, Blue, or Red interior decor. The Lycoming -L and -L1 models had "wet sump" crankcases of 12 qts. each, but the model -L2 with larger supercharged engines had an 8 gal. "dry sump" oil system.

It was noted that 19 of the Royal/Trecker "Gull" were still on the U.S. register as of 3-31-80; one was a model P.136-L, 12 were the model P.136-L1, and 6 were the model P.136-L2. The airplanes were divided about equally around the country. The Peruvian AirForce was reported to have 2, and a few examples were reported in Canada. Only one "Gull" was ever lost, and that was in Canada where flying can sometimes get very difficult.

ATC #814
(6-13-55)
VICKERS "VISCOUNT," V-700/800.

The Vickers "Viscount" put Capital Airlines back in the running.

The lovely four-engined Vickers "Viscount" was designed to take the place of older piston-engine types that were fast becoming obsolete on British short-haul and medium-haul airlines, routes where the new turbo-jet liners could not yet compete economically. The "Viscount" was designed around the new Rolls-Royce "Dart" kerosene-burning turbine engines which developed some 1400-1600 shaft-horsepower while driving huge four-bladed propellers; oddly enough, it was 2 Dart-powered Douglas DC-3 that were used by Rolls-Royce and Vickers for engine trials and development before the "Viscount" was introduced. The prototype "Viscount" (a V-630) first flew on 7-16-48 and received its British certificate of airworthiness on 9-15-49 becoming the first turbo-prop airliner designed and approved for scheduled airline service, and the first to go into production. The V-700 production version was modified from the prototype somewhat, and made its first flight on 8-28-50. British European Airways (BEA) placed the first order, an order for 20 of the V-701 type on 8-3-52 and deliveries began in Jan. of 1953; a month later BEA ordered 16 more. Meanwhile, the V-630 prototype had started the first ever scheduled turbo-prop service on a development basis from London to Paris in July of 1950; BEA started its scheduled service with the V-701 type in April of 1953. Before long the "Viscount" was winging away on routes to Cologne, Vienna, Zurich, Stockholm, Norway, Madrid, and then on to Istanbul, Cairo, and even as far away as Melbourne; all the while setting new records for speed on every route. "Viscount" service became so popular that BEA ordered 16 of the V-800 type to add to its fleet in 1956; by 4-22-57 the "Viscount" had carried its 2,000,000th passenger, and in less than a year later it had carried another million people! By this time the Vickers "Viscount" was in scattered use just about all over the world. British royalty flew in the "Viscount" frequently saying that they were comfortable in it, and trusted its ability; those that were not of royalty felt pretty much the same way. As the "Viscount" migrated to places like Africa, India, Australia, and Canada, it is logical to assume that some of the American air-carriers would become interested also. Capital Air Lines was the first American line to use the "Viscount," and with it they had made the rest of the industry sit up and take notice.

Having a unique problem on its far-flung routes scattered all over the eastern half of the U.S.A., and having to compete with the "big four" airlines for a lot of its business, Capital Air Lines (formed in 1948) began a search for an American transport that would give them an advantage, or even help them compete effectively. Because no American manufacturer was building this type of airplane, the boss-man at Capital went abroad to see what was available there; in England he found the near-perfect solution. What he found was the turbine-powered "Viscount" which would fit their needs to a tee; already proven in service the Vickers "Viscount" with its 4 turbo-prop engines was whisper-quiet, practically without vibration, it would be the right size, and above all it was economical for the type of service that "Capital" was rendering. The first 3 "Viscount" were ordered on 3 June 1954 and Capital got an option for 37 more; in August an option was taken for 20 more making a fleet total of 60 "Viscount" on order at a cost of $67,500,000!! Capital's first "Viscount" was christened into the fold on 23 June 1955, and its first flight on scheduled service was on 26 July 1955; this flight ushered in the "jet age" and heralded the invasion of a foreign transport airplane into an arena of heretofore American supremecy. The "Viscount" carried it off well. As additional "Viscount" were put into service (the V-745 was a version especially outfitted for Capital) the airline was progressively disposing of their various DC-3, DC-4, and Lockheed "Constellations." The four-engine "Viscount" was an immediate hit with the flying-public because it was smooth, quiet, fast, and comfortable; this was now giving Capital Air Lines an edge on routes where the competition had been the greatest. This new-found fame for Capital Air Lines led it into over-extending itself in routes and in equipment which eventually was causing financial problems. By 1957 things got worse instead of better, so Capital now just one step ahead of the sheriff, was seeking a merger with someone to forestall bankruptcy. Because of failure to meet payments, Vickers had taken back 15 of the "Viscount" from Capital by 1960 to sell to other lines; Capital's "Viscount" fleet was now down to 45. After overtures and many preliminaries, a merger with United Air Lines was consummated in June of 1961 and both lines came to benefit from the transaction. In the reshuffle the 41 inherited "Viscount" were all refurbished by United Air Lines to like-new condition. These and 6 more that "United" had bought back from Vickers were placed on short-haul routes where they wouldn't compete directly with the new turbo-jets that other lines were now beginning to use; thus the "Viscount's" service life was extended into 1969. By this time the "Viscount" had pretty much had its day in the limelight, and all were pulled from UAL service to be sold off to American business and to foreign air-carriers. A small item in a national

"Viscount" is readied for boarding.

"Viscount" 798 in service with Northeast Airlines.

magazine noted that 10 of the "Viscount" were donated to Embry-Riddle as classroom material.

The (British) Vickers "Viscount" V-700 type was a fairly large transport monoplane with seating normally arranged for 48 passengers and a crew of three. On the face of it the "Viscount" was a rather average-looking airplane, but it did harbor many innovations that were irresistable to passengers and operators alike. Its most prominent feature was the Rolls-Royce "Dart" turbine engines that drove huge propellers giving low fuel consumption of relatively cheap kerosene-type fuel; there was also about a 50% saving in engine weight, and the engines provided the efficiency of less aerodynamic interference over the wing because of their placement and shape. A bonus feature was the near-absence of engine noise with practically no vibration, a feature that passengers and crew found much to their liking. As powered with the 4 "Dart" turbine engines of some 1600-1800 shaft-horsepower each the eager "Viscount" was blessed with a relatively high performance that was ideal for the short-haul routes; take-offs and landings were relatively short and climb-out was fairly rapid, so time spent at terminals and time spent getting back to cruise altitude was far less than normal. From what has been said about this airplane it is easy to believe that flight characteristics were indeed very pleasant, it was quite easy to handle, the airplane was very stable and altogether the passengers were offered a very enjoyable trip, be it on a short hop to the next town, or a longer cross-country jaunt. The four-engined "Viscount" fairly bristled with safety features, and it was actually able to fly to safety on any 2 of its engines; operational safety was one of the prime considerations of its design. It is appropriate to stress again that the Vickers "Viscount" was the first foreign airplane to invade the U.S.A. with such gusto, and it was doing quite well on the crest of its wave, until the introduction of the smaller jet-airplanes which forced the proud "Viscount" to take a back seat. Retreating to a new position the "Viscount's" role was then changed and it continued serving in the field of big-business, on charter to romantic places, and on the lesser airlines scattered all over the world. Because of the accelerated progress in aviation, many proud airplanes had come to suffer this same fate at different times. The type certificate for the "Viscount" series was first issued (in U.S.A.) to the V-744 on 6-13-55, the V-745 was approved on 11-7-55, and the pop-

U.S. Steel had "Viscount" in their fleet.

ular V-745D was approved on 2-23-56. By 1959 approval for the 744 and 745 was cancelled because none of this type was registered in this country any more. It is probable that 70 or more of the "Viscount" were imported to the U.S.A. by Vickers-Armstrong Aircraft, Ltd. of Weybridge, Surrey, England. The "Viscount" was designed by G. R. Edwards and his team of experts with an assist on engine problems by Rolls-Royce; more than 400 of the "Viscount" were built in all the different series.

Listed below are specifications and performance data for the "Viscount" V-745 as powered with 4 Rolls-Royce "Dart" 510 engines rated 1365 shaft-horsepower at 20,000 ft. (1600 h.p. available for take-off); length overall 81'9"; height overall (at rudder) 27'9"; wing span 98'8"; wing chord 178" at root tapering to 53" at tip; total wing area 963 sq.ft.; airfoil NACA-63 Mod.; wt. empty 37,900 lbs.; useful load 22,500 lbs. plus; payload with 1680 gal. fuel, & 3 crew 12,250 lbs. (46 pass. & 4400 lbs. baggage-cargo); provisional gross wt. to 64,500 lbs.; max. allowable level-flight speed 330; cruising speed (.65 power) 324 at 20,000 ft.; landing speed (with flaps) 94; stall speed (no flaps) 114; take-off over 50 ft. barrier 3960 ft.; service ceiling 25,000 ft.; normal fuel cap. 1680 ga.; fuel cap. optional 2340 gal.; oil cap. 19.3 gal.; cruising range (.65 power at 20,000 ft.) using 290 gal. per hour was 5.5 hours; average price per airplane was approx. $1,200.000. The V-810 as approved on 4-22-58 was similar to the V-745D except the fuselage was 46" longer, had rectangular entry door, & a slight structural beef-up. Power was 4 "Dart" 525 engines rated 1585 shaft-horsepower at 20,000 ft.(1730 h.p. for take-off). Gross wt. to 69,000 lbs. for a capacity of 56 passengers and 5 crew; all else more or less the same.

Structurally the big "Viscount" was a low-winged cantilever monoplane of an all-metal stressed-skin construction carried out in the latest state of the art; some parts of the construction might even be considered innovations. The rounded fuselage was pressurized and air-conditioned giving passengers a comfortable sea level altitude inside the cabin while flying at 15,000 ft., and maintaining cabin temperatures from 65 to 80 degs. The main passenger cabin was 6'5" high in the center aisle and 9'5" wide for spacious seating 4 abreast, with 2 seats either side of the center aisle. A prominent feature were the large picture-windows that passengers so enjoyed; each window was also an escape hatch. Passenger entrance was to the rear on the left side, and crew entrance was up forward; the 745D of "United" had retractable entry-steps that pulled up into the fuselage. The 700 series had oval entry doors, and the 800 series had rectangular doors. There were 3 freight compts. and each had its own access door. The flight-deck up forward was more or less conventional with grouping provided for best control and operation. The huge cantilever wing of all-metal construction was built around a single main spar with wing ribs tieing into leading edge and trailing edge members; the wing was covered with stressed-skin "Alclad" metal sheet. Bag-type fuel tanks were mounted in the wing fore and aft of the main spar of each side of the fuselage; "slipper tanks" for extended range were optional. The "Viscount" mounted 4 smaller

"Viscount" on Maui in Hawaii, in island-to-island service.

"Viscount" drops gear on approach.

engines as an added safety factor; normal flight could be maintained on any 2 of its engines in emergency. The twin-wheeled retractable (hydraulic) tricycle landing gear of 23'10" tread used oil-spring shock struts with Dunlop or Goodyear wheels fitted with non-skid hydraulic brakes; the nosewheel was steerable. A thermal (hot air) de-icing system kept the wing and tail group free of ice by piping hot air from the engines thru' the leading edges. Exterior schemes and interior configurations were an option of the customer. The V-800 series were similar except for a stretched fuselage giving more capacity; the engines were also of higher horsepower. Each airplane had a complete set of airline equipment that included navigation & operational aids, four-bladed constant-speed, full-feathering "Rotol" props, a pressurization & air-conditioning system, a de-icing system and anti-icing system, double-slotted wing flaps, lavatory, pantry, snack-tray on the back of each seat, ash trays, hat & coat closet, overhead hand-bag racks, life-belt under each seat, 3-way adj. trimming tabs, safety lock for controls (can't take off with controls locked), and numerous extra-cost options.

It was noted that 27 of the Vickers "Viscount" were still listed on the U.S. register as of 3-31-80; 3 were the model 744, 23 were the 745D, and one was a model 810 owned by entertainer Ray Charles.

ATC #815
(5-4-56)
DE HAVILAND-CANADA "OTTER," DHC-3.

"Otter" was capable of high-performance with large payloads.

The smashing success of the popular DeHaviland "Beaver" (DHC-2) led to the design of a much larger airplane that was first called the "King Beaver"; it was later renamed the "Otter" so as not to encroach upon the well-liked "Beaver's" identity. Work on the prototype airplane, an airplane combining "Beaver" performance with a greater payload and longer range, was started in 1950 and its first flight came on 12-12-51 with George Neal at the controls. It was plain for everyone to see that here indeed was another winner. As a DHC-3 the "Otter" was approved both as landplane or seaplane, and was eligible to operate anywhere in the free world; its Certificate of Airworthiness was awarded just 10 months after its maiden flight. With double the payload and still having the same outstanding performance as the smaller "Beaver" the "Otter" soon picked up orders from Canada, So. America, and Europe; first delivery was to the RCAF in Nov. of 1952. Its tremendous versatility, lauded by bush-pilots, soon became known worldwide; then too, many of the "Otter" customers were already using the "Beaver," and production was hard put to keep up with the orders. An airplane which can meet the uninviting requirements of the Yukon and the Canadian North is bound to compel attention from other parts of the world, especially where airfields are a problem, where resources are scant, and a bare minimum of maintenance still requires months of reliable service. The big "Otter" had the built-in ability to overcome all these problems in its normal stride, and still deliver a performance that was a remarkable thing to behold. As the "Otter" were rolled out the assembly-room door they were most always sent to far-away places where the going was rough, places like the remote areas of New Guinea, Burma, Australia, Africa, etc.; here to operate in the steamy heat on wheels or floats, and even on amphibious gear out of places that would scare an ordinary airplane half to death! Designed to be adaptable to varied climates, some "Otter" operated in heat as high as 140 degs., and others operated in temperatures down to 60 degs. below zero; gads, what an airplane. Over 400 of the "Otter" were eventually built, and they were flying regularly in over 30 different countries, both in civil operation and in military service. In 1953

The big "Otter" as a seaplane, the bush-pilot's favorite.

the "Otter" astounded U. S. Army officials at Fort Bragg in comparative tests of fixed-wing airplanes against helicopters; the "Otter" proved that it could actually take off shorter with a much greater load, operate with more efficiency, and much more reliability on the same power. An order for 95 of the "Otter" was soon placed and first delivery was on 3-14-55 as the U-1, being the first airplane so designated in the new military category. In 1956 "DHC" was delivering the improved U-1A and by 1958 nearly 200 were already in service; many of the "Otter" saw war-time service in Viet Nam. A remarkable airplane such as the "Otter" will be hard to replace, so they are sure to be around, doing what they do best, for many years to come.

The big slab-sided (DHC-3) "Otter" by DeHaviland of Canada was an all-metal high-winged cabin monoplane, an airplane that was arranged quickly for a multitude of services in

"Otter" in U.S. Army as U-1.

The "Otter" in international service; it was a citizen of many nations.

areas where utility and ability were the foremost considerations. The large interior as specifically designed was quickly adaptable to varied density seating for up to 14 passengers, as an air-borne ambulance for up to 6 litters and 4 sitting patients, or cleared of all to carry 260 cu. ft. of all-cargo shipments. The quick change-over from job to job was perhaps one of its most popular features. Its ability to operate from all sorts of terrain on wheels, skis, or floats, on short notice, was a boon to operators too where the airplane was the only practical method of transportation. It is not unusual then that nearly 80% of the "Otter" built were exported to remote areas of the world. As powered with the reliable 9 cyl. Pratt & Whitney R-1340-H engine of 600 h.p. the "Otter" astounded everyone with its seemingly effortless performance in all kinds of service, and pilots reveled in its ability; its STOL characteristics (Short-Take-Off & Land) were hard to equal in an airplane of this type. Flight characteristics were quite pleasant and it did everything with a measured confidence; the "Otter" could slow-fly safely at 80 m.p.h., or step out at 140. Having no aerodynamic quirks the "Otter" could haul anything that was strapped-on without a fuss. and this included the carrying of canoes, bundles of lumber, or the attachment of spray equipment. Of course, pilots had to learn certain techniques to get the most out of an "Otter," but the airplane was so cooperative that it almost showed the pilot the way to do it. True, the big "Otter" borrowed much of its design from the smaller "Beaver," but the "Otter" was there to compliment and not to compete. If we note, each era had its outstanding airplanes, but one must concede that the "Beaver" and the "Otter" both spanned several eras. The type certificate for the DHC-3 "Otter" was issued 5-4-56 here in the U.S.A., and well over 200 units were imported into this country. Most tallies say that well over 400 units were mfgd. by DeHaviland Aircraft of Canada, Ltd, at Downsview, Ontario, Canada. P. C. Garratt was mng. dir.; W. D. Hunter was chf. of engrg.; C. H. Dickins was sales mgr.; W. Burlisson was prod. mgr. The "Twin-Otter," having 2 engines, first flew in 1963 to begin yet another cycle in the "Otter's" exciting history.

Men show relative size of the "Otter."

Listed below are specifications and performance data for the DHC-3 "Otter" as powered with 9 cyl. geared "Wasp" S1H1-G (R-1340) engine rated 550 h.p. at 2200 r.p.m. at 5000 ft.

(600 h.p. at 2250 r.p.m. for take-off); length overall 42'0"; height overall (at rudder) 10'6"; wing span 58'0"; wing chord 78"; total wing area 375 sq. ft.; wt. empty 4000 (4400) lbs.; useful load 3200 (3600) lbs.; payload with 214 gal. fuel 1774 (2163) lbs.; payload with 150 gal. fuel 2050 (2450) lbs.; gross wt. 7200 (8000) lbs.; normal gross wt. later raised to 7600 lbs. with a provisional gross wt. to 8000 lbs. with installation of stall-plates; max. speed 164 (155) at 5000 ft.; redlined at 192; cruising speed (.66 power) 140 (133) at 5000 ft.; economical cruising speed (.55 power at 5000 ft.) 130 (12); landing speed (with flaps) 54 (57); stall speed (no flaps) 64 (67); take-off run (over 50' barrier) 970 (1240) ft.; climb 1050 (990) ft. first min. at SL; land (over 50' barrier) 1000 (1090) ft.; ser. ceiling 18,000 ft.; gas cap. 214 (US) gal.; oil cap. 10.8 (US) gal.; cruising range (.66 power at 5000 ft.) using 35 gal. per hour was up to 6 hours; price in 1963 had gone up to $96,500. Seaplane performance was just slightly less.

The robust construction of the "Otter" was more or less comparable to that of the "Beaver" except that everything was of larger dimension. The floor was reinforced for heavy loads and the interior was serviceable, but nothing fancy. The cavernous cabin interior (12.5' x 5' x 5') for 260 cu. ft. capacity actually had room for 14-15 seats, and all were quickly removable to allow over a ton of cargo space with appropriate tie-downs. There was a large door each side for cabin loading, and a smaller door each side for cockpit entry; a baggage locker measured 27x-50x50 ins. for 51 cu. ft. capacity. Fuel cap. was 214 gal. in 3 tanks (62-102-50) mounted under the floor for easier servicing. The long and slender semi-cantilever wing was similar to that of the "Beaver" having slotted wing-flaps, and drooping ailerons that also contributed to slow-flight ability; stall-strips & stall-plates were a DHC modification that allowed an increase in the gross wt. No flaps above 94 m.p.h. The robust landing gear had a tread of 11'2"; skis and Edo 55-7170A floats were optional. Main wheels were normally 11.00x12 with hydraulic brakes, and the swiveling tail wheel was 6.00x6; amphibious float-gear was optional. The all-metal cantilever tail group was also a semi-monocoque structure; movable surfaces had aerodynamic balance and all these surfaces had adj. tabs for trim. A Hamilton-Standard constant-speed prop, engine starter, generator, 24V battery, oil cooler, misc. pumps, normal set of engine & flight instruments, compass, airspeed ind., cabin heaters, cabin vents, cabin lights, navig. lights, landing lights, & tie-down rings were std. equipment. Special interiors, long-range fuel tanks, DH skis, Edo pontoons, radio gear, canoe-carrying eqpt., lumber-carrying eqpt., crop spraying & water-dropping eqpt., ambulance interiors, & fire extinguisher were among the op-

The "Otter" on go-anywhere amphibious floats.

tions available. Seaplane required the addition of a ventral fin.

It was noted that 21 of the "Otter" were still on the U.S. register as of 3-31-80; 12 were registered in Alaska (ser. #54 as NC-3904 was with the Kachemak Air Service), and 11 airplanes of the overall total are still in some sort of government service. All were built in the period from 1955 to 1963.

Oversized wheels allowed landings on soft ground.

ATC #816
(6-24-57)
DE HAVILAND "HERON," DH-114

DeHaviland "Heron" with four "Gipsy" engines.

It is surely not intended as a slight when saying that the "Heron" was more or less an enlarged, four-engined version of the DeHaviland (DH-104) "Dove"; this because the popular "Dove," without a doubt, had proven itself to be quite an airplane. In the course of things the "Heron" (DH-114) was especially designed to serve where the smaller "Dove" was not adequate; on short-haul lines the "Heron" by comparison was able to carry up to 19 passengers, plus a crew of 2, with comparable reliability and economy. Using many "Dove" components in its makeup to lower manufacturing costs and speed-up construction, the four-engined "Heron" made its debut and maiden flight on 5-10-50 with Geoffrey Pike at the controls. Shakedown flights within England were started in the summer of 1951 and occasional cross-channel flights to Paris started soon afterward. Designed as an airplane that could be operated easily from the smaller airfields, and be maintained more easily where facilities were limited, the first 40 of the "Heron" had fixed tricycle landing gears. In 1952 the improved "Heron 2" was introduced with a retractable landing gear, more or less by demand, and it was 20 m.p.h. faster; operating abilities were still excellent and the extra maintenance was minimal. However, the "Heron 2" did not get into service until 1956.

A New Zealand airline bought the first production "Heron" and many orders followed; soon they were being shipped to India, Australia, Fiji Islands, Bahama Islands, W. Africa, and many places in between. Seeking to broaden their sphere of sales DeHaviland also developed a special "Executive" model (2DA) for the American market; these were generally plush and custom-built to order. Britishers remember the "Heron" as the last airplane to fly out of the famous Croydon Airport in London when it was shut down on 9-30-59. Tallies since show that nearly 150 of the "Heron" were delivered in civil and military versions to at least 30 different countries before production was discontinued in 1964. Perhaps the biggest single user of the four-engined "Heron" was the Puerto Rican International Airlines (Prinair) who started out with one "Heron" as an air-taxi service. Because of the good service offered the operation steadily expanded to nearby islands in the Caribbean Sea with 10 different stations, and became the world's largest commuter airline; the fleet had eventually grown to 22 of the "Heron," and not too long ago it stood at 29! The unusual "Heron"

"Heron" provided versatile service at a good profit.

was very popular and profitable for this type of service, but the 6 cyl. inline DeHaviland "Gipsy Queen" engines were aging and becoming a maintenance problem; the "Gipsy" engines were then run-out for the max. time allowed before a complete overhaul was necessary and then replaced one by one with 6 cyl. (opposed) Continental engines which were more powerful and easier to keep in good running order. Because the "Heron" was shipped all over the world, and flew long enough to wear out its "Gipsy" engines, it often wound up being powered with an interesting variety of engines, but in any case, it was still a very good airplane in just about any combination. The "Heron" was no show-off, but

British-marked "Heron" in flight on two engines.

it did pile up a very impressive operating record; there is really no foreseeable end just yet to its useful service.

The DeHaviland (DH-114) "Heron" was not an ordinary airplane because a small four-engined airliner not much bigger than the average "twin" was something you wouldn't see very often. However, when you did see it and saw it operate it always left a good impression. As a low-winged cabin monoplane the "Heron" was normally fitted to seat from 17 to 19 commuting passengers, but arrangements for the seating of 10, 13, or 15 passengers were also available according to the seating density required; the 17-19 passenger versions were allowed no lavatory. Designed especially for the small commuter line the "Heron" could go into and come out of the smaller airfields with four-engine safety, and could operate more cheaply for a better profit. The special executive version was naturally more plush and luxury interiors were arranged to the dictates of the customer; there was room for couches, office equipment, a pantry, a fully equipped lavatory, or just plenty of space to roam around or lounge in. But, the "Heron" was just as happy hauling commuters, or package-cargo. Another fact in its favor was its ability to operate in diverse conditions ranging from the Arctic to the Equator. The first "Heron" to go into service here in the U.S.A. was for the personal use of the British ambassador, and others soon followed to serve in the field of business. As powered with four DeHaviland "Gipsy Queen" engines of 250 h.p. each the "Heron" was surprisingly capable airplane, and the apparent four-engine safety was its attraction to passengers and crew alike. The pilots kept very busy with 4 engines to take care of, but the "Heron's" docile characteristics made the job a lot easier. As one pilot said "the most difficult thing about flying the "Heron" is learning how to get in and out of the cockpit properly"! One can't imagine that a four-engined airliner would be all that much fun to fly, but pilots have said it was and they enjoyed it; making up to 10 or 12 flights daily in commuter service a pilot would have to like it. The type certificate for the "Heron 2" series here in the U.S.A. was issued 6-24-57 with amendment on 5-19-58 for the deluxe model 2DA. Some 148 of the "Heron" (in all series) were mfgd. by the DeHaviland Aircraft Co., Ltd. on Hatfield Aerodrome in Herts, England.

Listed below are specifications and performance data for the (DH-114) "Heron 2A" as powered with 4 DH "Gipsy Queen" Mk. 2 engines rated 250 h.p. at 2500 r.p.m. at SL; length overall 48'6"; height overall (at rudder) 15'7"; wing span 71'6"; total wing area -499 sq.ft.; wt. empty 8480 lbs.; useful load 5020 lbs.; payload with 494 gal. fuel & 2 crew 1394 lbs. (6 pass. & 374 lbs. bag.); payload with 300 gal. fuel & 2 crew 2580 lbs. (13 pass. & 270 lbs. bag.); gross wt. 13,500 lbs.; max. speed 198; cruising speed (.65 power) 183 at 8000 ft.; landing speed (with flaps) 68; stall speed (no flaps) 75; climb 1410 ft. first min. at SL; ser. ceiling 18,500 ft.; max. gas cap. 494 (US) gal.; oil cap. 43 (US) gal.; max. cruising range (.65 power at 8000 ft.) using 60 gal. per hour was 1200 miles; price given only upon request.

The "Heron" was basically of all-metal (alum. alloy) semi-monocoque construction and used several "Dove" components in its overall makeup. The semi-monocoque fuselage was basically a "Dove" unit with an added section in the middle to accommodate the extra seating; there was a single row of seats each side of a center aisle. There was excellent visibility and everybody had a window seat. Normal seating was 2 crew and 13 passengers, but a (19-21 pl.) 17-19 passenger layout for air-taxi was optional. Baggage was under the cockpit floor and the entry door folded out & down to form steps into the cabin. The 2 outer wing panels were also "Dove" units with a large center-section added to accommodate the mounting of four engines. Four fuel tanks were in the wings, 2 on each side for a max. cap. of 494 (US) gal.; oil cap. was 43 (US) gal. in 4 nacelle tanks. The retractable landing gear was a pneumatic system operated by air pressure, as were the wheel brakes; the pneumatic system was clean and relatively trouble-free. The cantilever tail group was basically a "Dove" unit with enlarged surfaces and an assymetrical horizontal surface to eliminate buffeting. All movable controls were aerodynamically balanced; trim tabs were adj. for 3-way trim during flight. Normal interiors were fitted for commuter line service and air-taxi, but an ambulance interior, a cargo interior, or an executive version with plush surroundings for 6 people were optional. Engines & propellers

"Heron" was popular in the business-fleet.

were "DeHaviland," most of the landing gear & braking system was by Dunlop, & a full set of airline equipment was standard. The series 2A and 2DA had special furnishings for the American market.

It was noted that 47 of the "Heron" were still on the U.S. register as of 3-31-80; three were operating in Hawaii, 29 were with Puerto Rican International Airlines (Prinair) in the Caribbean area. The balance of the total (15) were pretty well scattered around the U.S.A. At least another 50 were still flying elsewhere around the world.

The "Heron" as modified with American engines.

TAG Airlines found "Heron" ideal for their service.

ATC #817
(10-29-57)
FOKKER "FRIENDSHIP," F-27

Fokker astonished the world with the "Friendship."

Seeking to design a suitable replacement for the aging Douglas DC-3, and also to regain its former foothold in worldwide commercial aviation, Fokker Aircraft had solicited ideas, hints, and requirements from airlines all over the world. As the questionnaires began to form a concept Fokker based this new design, the F-27, on their earlier F-24 high-winged monoplane which had shown some promise. The market study showed that a 40 passenger high-winged monoplane with tricycle landing gear that could operate from the smaller airfields at all altitudes, in weather ranging from very hot to very cold, would be ideal for the many smaller lines that served a large portion of the world at that time. As the piston-engine for aircraft had now reached its peak of development, and the pure-jet engine was still uneconomical for the shorter routes, it was decided to use 2 of the new Rolls-Royce turbo-prop (turbine) engines; these "Dart" engines as compared to piston-engines were much lighter, more economical, and had already been proven as very reliable in hard service. The high-wing configuration, as laid out by H. C. Van Meerten (Fokker's chief engineer), was chosen because it would offer a lower floor-level for easier loading, give the engines more clearance from the ground, but also because this type of ship was more stable and offered better all-around visibility. Because this ship would be used worldwide, hopefully, it was aptly named the "Friendship." Announced in 1953 as already being in development, the Fokker F-27 harbored several innovations in design and fabrication; the most novel of which was the bonding of metal to metal with adhesives instead of fasteners and rivets. The bonding method proved to be an ideal way of building strong, carefree, and efficient airplanes.

Three prototype airplanes were started in 1953, the first of which was flown on its maiden-flight on 11-24-55; the second prototype airplane made its maiden-flight in Jan. of 1957 and then joined airplane #1 in an extensive test and development program. The 3rd prototype, and a 4th airplane that was built up later, were used for all sorts of static testing. The very first order to Fokker Aircraft for an F-27 came on 3-9-56 from Trans-Australia Airlines, and by June some 13 airplanes had already been ordered. There was a little slow-down in orders during 1958, but interest in the airplane resumed again in 1959; because

Planform shows unusual "turbine" nacelles.

the Fokker (F-27)"Friendship" was such an ideal concept for service all over the world, the orders kept coming in continually. Most tallies show that some 475 or so of the F-27 "Friendship" were sold by Fokker Aircraft and it was still being marketed in the 1970's, altho' by now in a steadily-improved form. The Fokker "Friendship" has undoubtedly earned itself a revered place in aviation history as the most popular, and most successful twin turbo-prop airplane ever built! Here in the U. S. A., Fairchild Aviation had also made some studies for a design concept that would some day soon be needed as a "DC-3 replacement," and as their thinking more or less coincided with that of Fokker, they negotiated a build-under-license agreement. On 4-26-56 Fairchild signed the agreement to build the F-27 type under license, and watched Fokker's development program very closely. Reams of drawings and engineering data were soon sent to Fairchild so that tooling up could begin. Fairchild received their initial order for their version of the F-27 from West Coast Airlines for 4 airplanes in April of 1956, Bonanza Airlines ordered 3 in May, and Piedmont Airlines ordered 7 airplanes in June of the same year. From then on the orders kept coming in at a satisfying rate, and Fairchild began exploring the field of big-business with an "Executive" version. Fairchild's first delivery of a production version was in June of 1958 and a new era in short-haul air transport was begun in America by West Coast Airlines; in April of 1963 the 201st Fairchild F-27 was rolled out the door at Hagerstown, Md. It was an F-27F "Executive" model. After some 206 examples of the F-27 were built by Fairchild in several different versions it had proven itself by then to be a workhorse airplane thruout the western hemisphere, and truly a fitting successor to the time-honored DC-3, if there ever can be such a thing. The American-version of the Fokker F-27 was built by the Fairchild-Hiller Corp. at Hagerstown, Md. under Type Certificate (TC) 7A1 that was first issued 11-13-57 and amended at different times to approve the various versions they had built. Fokker and Fairchild must have had a very congenial agreement between them because they were actually delving in one another's markets, perhaps because of customer preference; some in the western hemisphere bought Fokker aircraft and some Europeans bought Fairchild airplanes.

The Fokker "Friendship" (F-27), seen at just about every air-terminal around the world, was an all-metal high-winged transport monoplane with 2 "turbo-prop" turbine engines; it was arranged to seat anywhere from 32 to 56 passengers, depending on the interior configuration. Standing tall with its belly close to the ground it was but a short step into the airplane, more like boarding a bus; some versions were equipped to carry some cargo along with the passenger load, while some were fitted to carry cargo only. One of the big reasons for success of the "Friendship" was its exceptional versatility; it could be developed or adapted for a sheer multitude of tasks in both civilian and military roles. Being able to operate from major airports or unpaved strips at various altitudes in all kinds of weather, it soon became tops in its role as a medium-sized transport. By virtue of design the maintenance was minimal, and when need be the F-27 could be tended to as easily on some desert airstrip as in the most sophisticated hangar; replacement parts were easy to get and shipment from the factory was prompt. Because of built-in reliability and the continuous upgrading of structure and accessories, as time went on, major maintenance was not generally required until 20,000 or more hours. On such attributes the "Friendship's" reputation was built.

As powered by 2 Rolls-Royce "Dart" turbo-prop (turbine) engines rated from 1535 to some 1835 shaft-horsepower the F-27 in all its versions had an eager and deft performance that served it admirably in all kinds of use and service. The flight characteristics were simply excellent which allowed pilots to do things with the craft that were unbelieveable for an airplane of this type. It has many times been said that the "Friendship" (F-27) was very well liked, and even loved, by everyone ranging from the front-office, the operating crew, and right down to the mechanics that had to work on her. Of course, the paying passengers usually had nothing but praise for the time they were on board. The famous DC-3, by the way, was also accorded such affection and it is then fitting that the Fokker F-27 be thusly blessed. Fairchild discontinued manufacture of the F-27 series in 1963 and gave up their licensing agreement in July of 1973. The F-27 by Fokker was planned to remain in production for the foreseeable future, and who knows when it will end! The "Friendship" is sure to be around for several more years. Application for an American "import" Type Certificate had been made by Fokker Aircraft in 1954 at a meeting in Washington, D.C. with the CAA; it was the first such meeting ever held here for the certification of a foreign-built airplane. The type certificate for the "Friendship" F-27 was first issued on 10-

An F-27 ready for boarding.

Fairchild built "Friendship" under license, it proved to be a hustler.

29-57 for the Mark 100, and approval for the Mark 200, 300, 400, 600, 700 versions was awarded on 5-25-65; approval for the Mark 500 with stretched fuselage came on 5-21-70. The "Friendship" in all its versions was mfgd. by the Fokker Aircraft Co. of Schiphol, The Netherlands; they did have several plants, and sub-contractors also mfgd. various sub-assemblies. Fokker built 475 units to 1978 and Fairchild built 206 units to the cut-off date in 1973 for a total of 681 airplanes mfgd. to 1978.

Listed below are specifications and performance data for the Fokker-built F-27 "Friendship" (Mark 100) as powered with 2 Rolls-Royce "Dart" Mk. 511 turbine engines rated 1400 h.p. at 14,200 r.p.m. (1570 h.p. at 14,500 r,p.m. with water injection for take-off); length overall 75'9"; height overall (at rudder) 27'11"; wing span 95'2"; wing chord 11'4" at root tapering to 4'7" at tip; total wing area 754 sq. ft.; airfoil NACA-644-421 Mod. at root tapering to NACA-642-415 Mod. at tip; wt. empty 24,838 lbs.; useful load 15,662 lbs.; payload with 972 gal. fuel & 2 pilots 9930 lbs. (40 pass. & 2620 lbs. baggage-cargo); provisional gross wt. to 40,500 lbs.; empty wts. & gross wts. were later raised for subsequent versions; max. speed approx. 280; max. cruising speed 259 at 10,000 ft.; landing speed (with flaps) 75; climb 1640 ft. first min. at SL; ser. ceiling 25,000 ft.; fuel cap. 972 gal.; oil cap. 8 gal.; water/methanol mixture 80 gal.; cruising range (using 233 gal. per hour) was 1000 miles; price was only given upon inquiry. Fuel cap. was optional and more powerful engines were used on subsequent versions.

The Fokker F-27 was of all-metal construction with a semi-monocoque fuselage and a cantilever stressed-skin wing of very high aspect-ratio; the tail-group was also an all-metal cantilever structure. Materials largely used were 7075, 7079 & 2024 aluminum alloys both bare and "clad." Chrome-moly steel, stainless steel, and magnesium were also used in various components. Fiberglass was used for ducting, window framing, paneling, and for other secondary structures not needing structural strength in the airframe. Metal to metal resin bonding was used for

"Friendship" was swift, beautiful, and very efficient.

attachment of much of the "skin," and stringers. The fuselage of 100" diameter was pressurized for almost its entire length, and air-conditioned for cabin comfort. A passenger door for entrance was to the rear on the left side; the crew had a separate entry door up forward, and a large cargo door up front was optional. Cargo-planes had a reinforced floor; large cargo doors were in front, and in the rear too if needed. Normal passenger seating was 4 abreast with an aisle down the center; the toilet, lavatory, and the pantry were to the rear of the cabin area. Normal airline seating was 40 passengers, but various other arrangements were optional. The Mark 100, 200, 300, 400, 600, 700 had max. seating to 48 passengers; the Mk. 500 was same as the 600, but had stretched fuselage for seating of up to 60 passengers. The cabin was pleasantly quiet, very comfortable, and visibility was excellent from any seat; and too, there was practically no vibration. A fuel cell was mounted in the wing either side of the built-in center-section; pylon tanks for extra range were optional. The outboard wing panels were removable at the engine nacelles. The retractable tricycle landing gear of 23'7" tread had dual Dunlop wheels all around and was retracted or extended by air-pressure; the air-type retraction system was light, clean, and fireproof. The 27" nosewheel was steerable; a special rough-field landing gear was optional. The wheel brakes were also operated by air. No wing flaps down above 140 m.p.h., and no landing gear down above 170 m.p.h. The cantilever tail group had adj. trimming tabs as did the ailerons. The F-27 flew very well on only one engine, and could actually climb out loaded on one engine in emergency. Dowty-Rotol constant-speed full-feathering props, a Roots-type pressure system providing an 8000 ft. altitude in the cabin at 25,000 ft., air-conditioning system that was active on the ground as well as in the air, a de-icing & anti-icing system, a fire extinguishing system, emergency oxygen system for pilots, and a complete set of airline equipment was standard. Many options were available at extra cost.

It was noted that 12 of the Fokker-built (F-27) "Friendship" were still on the U.S. register as of 3-31-80. All were bearing U.S. registration numbers, and 7 were with the Arabian-American Oil Co. It is of interest to mention that the Fairchild-built F-27 was approved on Type Certificate 7A1 first issued 11-13-57 with various amendments issued to cover the subsequent models to 6-12-69 for the model F-27M.

Intimate view of F-27 details.

BIBLIOGRAPHY

BOOKS

Airway One by Robt. E. Johnson
Boeing Aircraft Since 1916 by Peter M. Bowers
British Civil Aircraft by A. J. Jackson
Cessna Guidebook by Mayborn & Pickett
DeHaviland Aircraft Since 1915 by A. J. Jackson
High Horizons by Frank J. Taylor
Pedigree of Champions by The Boeing Co.

MAGAZINES

Flying
AOPA Pilot
Aero Digest
Air Classics
Western Flying
Armchair Aviator
Rotor-Craft
Air Progress
Sport Flying
Plane & Pilot
Private Pilot
Journal of AAHS

SPECIAL MATERIAL

Airline company data, factory data, government data, and correspondence with individuals.

PHOTO CREDITS:
for ATC Section . . . (including alternates) (60-13)

Balzer, Gerald Collection, 20, 26, 31, 59, 60, 62, 63
Bowers, Peter Collection, 5, 23, 24, 29, 48, 50, 68
Fairchild-Hiller Corp., 25, 71, 72, 73, 76
Juptner, Jos. P. Collection, 9
Koston, Ted (courtesy of EAA), 15
Larkins, Wm. T., 6, 7, 18, 19, 25, 30, 36, 55, 56, 69
Molson, Ken Collection, 13, 34, 36, 38, 40, 64, 65, 66, 70
Pan American World Airways, 43, 45
Rare Birds, 12, 14
Smithsonian Institution, 22, 58
Underwood, John W. Collection, 26, 33, 49, 51, 54, 61, 67, 74
United Air Lines, 41, 42, 44, 46, 52, 53, 57
Williams, Edward D., 47
Williams, Gordon S., 1, 2, 3, 4, 8, 10, 11, 16, 17, 28, 35, 37
Wright, Jack Collection, 21

ATC UPDATE

In this section we hope to correct whatever errors had creeped up in the telling, of omissions that should have been included, and of "typos" that got by the proof-reader. Several chapters lack the proper continuity in a particular paragraph, so the correction is entered here as it should read.

Furthermore, to invite use of this section, it is liberally embellished with some interesting photographs that are pertinent to a certain airplane's development, or to point out some special use the airplane was involved in. The design, manufacture, and capabilities of an airplane is in itself an interesting study, but the diversified use of an airplane is sometimes also a very interesting story.

ATC #3

Unusual view of Johnson "Twin-Sixty" as modified into a cabin airplane.

ATC #8

Serpent-design paint job on early "Eaglerock" was used to promote dining in a Chinese restaurant.

ATC #14

Douglas C-1 transport refueling the famous "Question Mark."

ATC #17

1928 "American Eagle" (four-aileron) as a company ship; an early example of airplanes in business.

ATC #24

Testing early "Detroiter" (SB-1) biplane on floats in Lake St. Clair, Mich.

ATC #25

Early "Brougham" with "Hisso" engine; the water-cooled V-8 engine was a lot cheaper than Wright "Whirlwind."

ATC #32

Name in lower L.H. column of page 95 refers to Wallace Beery (not Berry), the famous movie-actor.

ATC #36

OX-5 "Pheasant" that "Steve" Wittman owned for many years; view is in front of Wittman hangar.

ATC #37

Pilots accepting delivery of rare "Model 7000" from Walter Beech; ship was to be used in Alaska.

ATC #39

Berliner "Parasol" showing protected accommodations for the passengers.

ATC #40

"Casey" Jones flying the original Curtiss "Robin" on its maiden flight.

ATC #42

A well-worn "Hisso-Waco" still going strong after a decade of service.

ATC #43

An all-revealing view of the Simplex "Red Arrow."

ATC #53

A more detailed view of the "Command-Aire" 3C3.

ATC #55

An exciting view of a Stearman C3B taking off; "Deed" Levy, company test-pilot flying.

ATC #58

"Eaglerock" A-2 (OX-5) off on a training flight; scene was in Los Angeles.

ATC #62

A brand-new "Hisso-Stearman" as it was rolled out at the factory in Wichita. Very rare.

ATC #90

Loening C2C meets the "Greyhound" bus at the Cleveland dock.

ATC #93

This ATC number also for the model 5-A "Executive."

ATC #95

The Mohawk "Pinto" MLV being tested on skis.

ATC #97

Rare photo of K-R "Challenger" C-2 with 7 cyl. Hallett engine.

ATC #98

Buhl "Senior Airsedan" (CA-8A) of 1931; probably the last one built.

ATC #113

Young lady in Fig. 35 is Florence "Flo" Klingensmith, a very lovely and accomplished aviatrix.

ATC # 124

A small folding-wing biplane "American Eagle" had hoped to market.

ATC #132

Rare view of Ford "Tri-Motor" 11-AT with three Packard Diesel DR-980 engines of 225 h.p. each.

ATC #137

Rare view of Stearman C2H with 260 h.p. Menasco-Salmson radial engine. (Rare).

ATC #139

Rare view of "Eaglerock" with 7 cyl. Hallett engine.

ATC #154

Fig. 176 shows International F-18 entry as modified for Dole Derby race to Honolulu; Frank Clarke pilot.

ATC #161

Photo in Fig 198 was credited to Peter M. Bowers, but actually taken in Alhambra by Boardman C. Reed.

ATC #163

The unusual Buhl "Autogiro" with pusher engine; only 1 was built under license to Autogiro Co. of America.

ATC #169

The six-place Lockheed "Vega" as used by Pan Am-Grace; no photo of this airplane was shown in Vol. 2.

ATC #184

Center-section fuel tank as shown in Fig. 272 was simply a cut-down 50 gal. oil drum.

ATC #187

Rare view of four-place Stearman "Coach" that was not marketed; Lloyd Stearman on left and Mac Short with cap.

ATC #202

Golden Eagle "Chief" in Canada; the airplane actually flew better without the NACA-type engine cowl.

ATC #208

The rare Fairchild KR-34-A with Curtiss "Challenger" engine; a photo was not shown in Vol. 3.

ATC #210

"Aristocrat" 102-E as traveling show-case for aircraft hardware; operated by Air Associates.

ATC #214

Striking in-flight view of the "Command-Aire" 5C3-B.

ATC #215

Fairchild KR-21-A as used on skis during New York winters.

ATC #222

Fokker F-11-A as test-bed with engine as a "tractor" installation instead of "pusher."

ATC #247

Doyle "Oriole" being tested with 120 h.p. Chevrolair-Martin 333 engine.

ATC #248

Big Bear and Arrowhead lakes should read as being in the San Bernardino Mountains.

ATC #262

Ryan B-7 as shown in Fig. 199 was operated by National Airlines.

ATC #265

"Cabinaire" prototype was built up by using most of the components from Walter Carr's own "Travel Air"!

ATC #283

"Phaeton" using new 190 h.p. Kinner engine on test; being flown here by Jean LaRene.

ATC #288

Eastman "Sea Rover" in Detroit's "Belle Isle" lagoon.

ATC #293

Rare "Porterfield Trainer" fashioned from an American Eagle model 201.

ATC #300

"Lindy" installs big "Cyclone" engine in his "Sirius" for survey-flights around the world for Pan Am Airways.

ATC #318

The original Alexander "Bullet" had 10 cyl. (French) Anzani engine for test.

ATC #319

Ship shown in Fig. 79 was on APC floats made in Canada and not on Edo floats as stated.

ATC #320

A "Fleetster" 20 working for Aerovias Centrales in Mexico.

ATC #329

Ship shown in Fig. 115 is not an SM-7B, but rather a "Model W" as covered by ATC #435.

ATC #337

Pleasing view of the Waco QSO as covered by ATC #337.

ATC #341

Experimental "Swallow" monoplane as designed by Dan Lake, shows a marked resemblance to Alexander "Bullet."

ATC #359

The Monocoupe 125 with 125 h.p. Kinner B5 engine; a photo was not shown in Vol. 4.

ATC #373

Caption for Fig. 224 should read model AP-K5, and not AP-B5 as shown.

ATC #377

Shell Oil crew refueling "Jimmie" Doolittle's Laird "Super Solution"; note the crude refueling gear.

ATC #405

Bottom of L.H. column on page 20 it should read that landing gear used rubber shock-cord; oleo-spring units were a modification. Tail skid was a steel tube snubbed with rubber shock-cord; spring-leaf skid was also a modification.

ATC #409

Ford "Tri-Motor" 5-AT-D (RR5) as used by U.S. Marines.

ATC #439

"Ole" Fahlin testing one of his "props" on an early 4 cyl. Continental A-40 engine.

ATC #451

Rare view of Northrop "Beta" sportplane with Menasco inline engine.

ATC #508

Ill-fated "Orion/Explorer" combination that took Wiley Post & Will Rogers to their death.

ATC #522

Ship shown in Fig. 71 was later modified by Wally Timm into the 2SA "Aerocraft" as covered by ATC #733.

ATC #593

French S-43 (Sikorsky) being refueled "somewhere in the tropics" from a motor launch; the French used the S-43 in the French colonies of Africa.

ATC #604

Young lady shown in Fig. 21 is Blanche Noyes, and not Louise Thaden as captioned.

ATC #609

SR-8EM (CF-AZI) shown here in Canadian service; see page 39 in Vol. 7 of same airplane (factory fresh) poised for delivery flight.

ATC #619

A bevy of stewardesses that graced the flights of "UAL."

ATC #625

In the registration listing on page 95, CF-BEA, -BEB, -BEC were ser. #5221-22-23 respectively. Credit photo of CF-BEB to Gordon Irons.

ATC #626

Plane shown on page 98 is a YKS-6 and not a ZKS-7 as captioned. Plane shown on page 99 is a VKS-7 and not a ZKS-7 as captioned. (Courtesy Ray Brandly)

ATC #628

Spartan "Executive" flown by Arlene Davis (page 106) was actually in 1939 National Air Races. Registration listing should read "ser. #10 in Bendix Trophy Race."

ATC #630

Prototype of the Beech 18 "Twin" shown here on one of its early test flights.

ATC #637

The little "Falcon" monoplane which was actually the first Welch OW-5; -11382 was ser. #107 registered to Orrin Welch. Info & photo courtesy Drina Welch Abel.

ATC #639

The Waco EGC-7 as ordered by Brazil; an EGC-7 was not shown in Vol. 7. The plane shown on page 139 of Vol. 7 is an EQC-6 and not an EGC-7 as captioned. Info & photo courtesy Ray Brandly of Nat. Waco Club.

ATC #643

The legend below the registration listing on page 156 states there was no listing for ser. #192; actually, ser. #192 was registered in Canada & is shown on page 155 as CF-BGR. By the way, CF-BGR is well & still flying.

ATC #646

For a more accurate & comprehensive write-up on the mysterious 7-WP refer to Peter Berry article in Summer 1980 issue (Vol. 25-2-145) of the American Aviation Historical Society Journal. Apparently it was a prototype that was exported to China.

ATC #647

On page 167 starting in lower L.H. column is a hopeless "scramble" that is not in proper continuity. In the lower paragraph on construction details it should read "As a transcontinental sleeper-plane the DSTA-SB3G was fitted with up to 14 berths (7 upper & 7 lower) in a cabin area that was 19'6" long x 7'8" wide x 6'4" high. The front sleeping compartment was often converted into a private 'Sky Room' with isolation for 2, complete with private lavatory. On day-time flights this compartment was often converted into a smoking lounge for 4 people."

ATC #665

This is the Waco EGC-8 "Custom Cabin"; a photo of the EGC-8 was not shown in Vol. 7. The photo shown on page 226 was actually an EGC-7 and should have been shown on page 139. Info & photo courtesy of Ray Brandly.

ATC #673

The last sentence in the first paragraph on page 253 regarding the 14-WG3B should read "Business and the profits of war often take some peculiar turns." Refer to Vol. 7.

ATC #714

A striking view of the "Waco E"—King of the "Waco-Cabin" line.

ATC #718

The prewar "Ercoupe" as a seaplane on Edo twin-float gear; shown here over New York City.

ATC #722

Cessna T-50 "Twin" as a feeder-line transport in early post-war years.

ATC #725

The war-time Piper AE-1 ambulance-plane ready to load a litter-patient.

ATC #728

The Aeronca 65-TAC as fitted with Heath seaplane gear.

ATC #732

This approval also for Ross "Parasol" model RS-2L with 50 h.p. Lycoming engine.

ATC #736

Annual home-coming of Meyers OTW to factory in Tecumseh, Mich. Planes came from all over the country.

ATC #743

An ex-Navy N2S-3 (post-war modification) on the job as a crop-sprayer over the "delta country" of Northern California.

ATC #746

Taylorcraft L-2B used in winter on skis. Correction: wings of the civil "Tandem," 4 of the YO-57, & first 20 of the O-57 all had fabric-covered all-metal wings; the O-57A, L-2A, L-2B, L-2M all had wooden wings covered with fabric.

ATC #752

Correction on page 183; the "Excalibur" did not crash in England, the tragedy happened in the Bay of Exploits in Botwood, Newfoundland.

ATC #753

The prewar "Swift" GC-1 warming up for one of its early flights. A photo of the GC-1 was not shown in Vol. 8.

ATC #756

The "Conestoga" as used by the "Flying Tiger Line." Correction: wording in the first paragraph should actually read "the first large airplane to be built of stainless steel.

ATC #758

A 1949 "Call-Air" on skis; note location of baggage compartment.

ATC #759

A "Champ" doing nip-ups at an air-show.

ATC #760

A rare view of the Piper L-14 in flight.

ATC #764

Stinson "Sentinel" on floats with U.S. Navy as OY-1.

ATC #767

Specification paragraph in lower L.H. column on page 235 should start out by reading "Listed below are specifications and performance data . . . etc."

ATC #768

Cessna 140 seaplane at rest in a sheltered lagoon. The 140A shown on page 240 was approved on TC 5A2.

ATC #769

The prototype "Seabee" (RC-1) had unbraced cantilever wing. P.H. Spencer shown in cockpit.

ATC #772

Slick Airways was the holder of this Type Certificate for remodeling the C-46A, D, F models.

ATC #775

The Taylorcraft Model 15 "Foursome"; a photo of this airplane was not shown in Vol. 8. The ship shown on page 260 is the Model 18 built later.

ATC #778

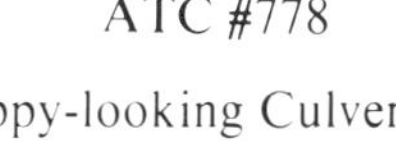

A snappy-looking Culver V-2 cruising along at over two miles per minute.

ATC #780

Piper "Super Cruiser" (PA-12) on Edo floats.

ATC #785

Chas. H. Babb held Type Certificate for model 28-5ACF.

ATC #787

A post-war "Ercoupe" up among the fluffy clouds.

ATC #788

Correction in L.H. column of page 305: it was J. O. "Bob" Noury and not Robt. Noury who designed the "Noranda." Noury's 2nd airplane became the "Canuck" and not the "Noranda" which was tandem-seated.

ATC #791

A pleasing view of the Thorp "Sky Shooter."

ATC #792

A classic view of the All-American "Ensign."

PHOTO CREDITS
for
ATC UPDATE

Abel, Drina Welch Collection, 92
Balzer, Gerald Collection, 77, 79, 80, 81, 96, 97, 99
Barnes, Monty Photo, 92
Beech Aircraft Co., 78
Besecker, Roger, 79
Bowers, Peter Collection, 85, 95, 97
Bowers, Peter M. 86
Brandly, Ray Collection, 92, 93
Call Aircraft Co., 97
Camart, Inc. (Vic Stein), 100
Cresswell Photo, 82, 86
Edo Corp. (Hans Groenhoff), 94
Engrg. Research Corp., 100
Fairchild Hiller Corp., 81, 84
Havelaar, Marion Collection, 77
Hudek Collection, 78, 82, 85, 86
Inman Co., 100
Institute of Aero. Sciences, 78
Juptner, Jos. P. Collection, 80, 90
Larkins, Wm. T., 96
Lebrecht, Chas., 87
Lockheed Aircraft Co., 90
Lowry, Louis M., 85
McCormick Armstrong Photo, 99
McVickar, F. C., 89
Meyers, Allen H., 85
Molson, Ken Collection, 91
Northrop Corp., 90
Panagra via R. S. Allen, 83
Payette Collection, 77, 83, 95
Pratt & Whitney Div., 91
Rare Birds Collection, 98
Shell Oil Co., 89
Smithsonian Institution, 77, 78, 79, 87, 88
Stearman Aircraft Co., 80, 82, 84
Steele, Bob Collection, 98
Trask, Chas. N. Collection, 94, 95, 96, 97, 98, 99
Underwood, John W. Collection, 84, 87, 89
United Airlines, 91
Williams, Gordon S., 79, 81, 88
Wilson, Kenneth D. Collection, 88, 94
Wittman, S. J. "Steve", 78

THE GROUP 2 APPROVAL

This section is a capsulized listing of all airplanes that were built under the so-called "Group 2 Approval." The listing here is in numerical sequence, and photos are shown wherever possible. This listing is designed too to be interrelated with the "ATC series" to present a better picture of airplane development in the period covered. We do not propose to have the following listings be a complete technical description and tell all, but instead we hope to whet the appetite for further research and help promote the gain of more knowledge.

The sometimes mysterious "Group 2 Approval" was awarded as a "Memo Number" or a "Letter of Approval," and was initially offered as a cheaper way to go for the small operation, or individuals, who could not see fit to spend the rather large amounts of money necessary for acquisition of a full-blown "ATC" (Approved Type Certificate). This "lesser" type of approval (Group 2) was usually awarded to an airplane that would be built either in one or two examples only, a limited quantity for test and evaluation, or for some predetermined number of airplanes; this approval was also awarded for certain modifications (such as a different engine, increased, decreased, or rearranged seating, major interior changes, increased fuel capacity, etc.) of a standard type airplane already in approved (ATC) production.

Some people in the business sort of looked down upon a "Group 2 Approval" as something far less than the vaunted "ATC," but actually, it was quite possible for a "Group 2 airplane" to be as good or even better-built by comparison; this because of the numerous mandatory inspections required for each stage of manufacture for each and every airplane built under Group 2 requirements. And, each and every airplane had to pass a satisfactory flight-test performed or witnessed by the government inspector before being approved. But then sometimes, if the inspectors were extremely busy and could not get to the site, some was done by paper-work and builder's affidavits to prove soundness of design and construction; perhaps this is where the doubts lay. The all-important phrase "when built to conform" was the key wording in any "approval." Experience has shown that an airplane awarded a "Group 2 Approval" was necessarily of sound design, of sound manufacture, and as airworthy as it was possible to make it. Some of our most interesting airplanes were built under the "Group 2 Approval."

#2-1 (1-1-29)

Alexander "Eaglerock" A-7 as 3POLB with 128 h.p. Ryan-Siemens (Siemens-Halske SH-12) engine; plane typical of the A-series. For ser. #451 (C-4570). Approval was cancelled because only airplane eligible was completely destroyed in hangar fire at Von Hoffman School of Flying.

Special Memo #2-1

Reissued to Alexander "Eaglerock" A-12 Spl. as a 3POLB with 165 h.p. Comet 7-E engine; for ser. #312 (C-3157) only. See ATC #139 for similar examples as approved later.

#2-2 (1-7-29)

Fokker "Tri-Motor" F-10 as 14PCLM with three 450 h.p. Wasp C-1 engines; this approval for first 5 airplanes as ser. #1000 thru 1004 at 12,500 lbs. gross wt. Refer to ATC #56 (Vol. 1-147) for photo of this early example.

#2-3 (5-22-29)

Fokker "Super Universal" as 6PCLM with 420 h.p. Wasp C engine; for ser. #800 (NC-3318) only at 4500 lbs. gross wt. This the first "Super Universal" had high-altitude wing. Refer to ATC #52 for subsequent examples.

#2-4 (10-22-28)

Boeing Tri-Motor 80 as 14 PCLB with three 450 h.p. Wasp C-1 engines; for ser. #1030 thru 1033 at 15,660 lbs. gross wt. See ATC #206 (Vol. 3-25) for photo of this type.

#2-5 (6-1-28)

Breese "Model 5" as 5PCLM with 220 h.p. Wright J5 engine; typical of the famous Dole Derby "Aloha." This approval for all similar airplanes at 3100 lbs. gross wt. Used on several early airlines in the west.

#2-6 (3-1-29)

Buhl "Army Trainer" as 2POLB with 220 h.p. Wright J5 engine; this approval for one airplane (X-1673) only at 2583 lbs. gross wt. Basically a modified Buhl "Airster"; refer to ATC #1 for basic design.

#2-7 (12-16-28)

Cessna BW as 4PCLM with 220 h.p. Wright J5 engine; this was a high-performance version of the Cessna A series. For ser. #113, 116, 117, 118, 120, 121, 125, 135, 138, 142, 143, 144, 147 at 2435 lbs. gross wt. See ATC #65 for more info.

#2-8 (1-5-29)

Cessna AS as 4PCLM with 128 h.p. Ryan-Siemens (Siemens-Halske SH-12) engine; typical of the A-series. This approval for all similar airplanes at 2260 lbs. gross wt. See ATC #65 for more info.

#2-9 (9-25-28)

Ford-Stout "Tri-Motor" 4-AT-A as 14PCLM with three 220 h.p. Wright J5 engines; for ser. #1 thru 14 at 9300 lbs. gross wt. Refer to ATC #87 for subsequent examples. Ship shown was used as "air truck" to deliver "Royal" typewriters.

#2-10

This approval first issued to Ford-Stout "Tri-Motor" 4-AT-B, but was superseded by ATC #87 (See Vol. 1).

#2-11 (10-27-28)

Ford-Stout "Tri-Motor" 4-AT-C as 14PCLM with 450 h.p. Wasp engine in nose & two 220 h.p. Wright J5 engines in the wings; for ser. #4-AT-47 (NC-7862) only at 10,000 lbs. gross wt. Refer to ATC #87 for other 4-AT series.

#2-12 (2-16-29)

Ford-Stout "Tri-Motor" 5-AT-B as 14PCLM with three 420 h.p. Wasp C engines; for ser. #5-AT-4 and up at 12,650 lbs. gross wt. if built for NAT and TAT. Refer to ATC #156 for similar examples.

#2-13 (8-27-28)

Hamilton Metalplane H-43 as 8PCLM with 420 h.p. Wasp engine; for ser. #43 only at 6000 lbs. gross wt. Also listed as H-21 when registered to Scenic Airways. Refer to ATC #85 for more info.

#2-14

This approval first issued to Hamilton Metalplane H-47, but was superseded by ATC #94. (See Vol. 1).

#2-15 (3-1-29)

International (Fisk) F-18 as 5PO/CLB with 220 h.p. Wright J5 engine; for ser. #11 and 12 at 4000 lbs. gross wt. Prototype of this version had 180 h.p. Hispano-Suiza V-8 water-cooled engine. Refer to ATC #154 (Vol. 2-156) for photo of "Dole Derby" version of this airplane.

#2-16 (11-16-28)

Ireland "Neptune" N-2 as 5POAmB with 220 h.p. Wright J5 engine; for ser. #16 and up at 3620 lbs. gross wt. Refer to ATC #153 (Vol. 2-153) for photo of (NC-6813) as ser. #17.

#2-17 (8-27-28)

Laird "Commercial" LCB as 3POLB with 220 h.p. Wright J5 engine; for all LCB mfgd. before 10-1-28. Eligible at 2850 lbs. gross wt. Refer to ATC #86 for more info.

#2-18 (7-18-28)

Mahoney-Ryan B-1 as 5PCL/SM with 220 h.p. Wright J5 engine; for ser. #61 only as seaplane on Fairchild floats at 3300 lbs. gross wt. Refer to ATC #25.

#2-19 (9-22-28)

Metal Aircraft "Flamingo" G-1 as 5PCLM with 450 h.p. Wasp C engine; for ser. #1 (C-7690) at 5000 lbs. gross wt. Refer to ATC #192 for more details.

#2-20 (11-1-27)

Pitcairn "Orowing" PA-3 as 3POLB with 90 h.p. Curtiss OX-5 engine; for all ser. nos. to #32 at 1965 lbs. gross wt. Refer to ATC #18 (Vol. 1-59) for photo of this type.

#2-21 (11-1-27)

Pitcairn "Fleetwing" PA-4 as 3POLB with 90 h.p. Curtiss OX-5 engine; for all ser. nos. to #7 at 1880 lbs. gross wt. Refer to ATC #18 for more info.

#2-22

Pitcairn "Super Mailwing"; approval superseded by ATC #92.

#2-23 (6-15-28)

Sikorsky UN-4 as 2POLM with 90 h.p. Curtiss OX-5 engine; a parasol monoplane with JN-4D fuselage with Sikorsky hi-lift wing. For all ser. nos. at 2100 lbs. gross wt.

#2-24 (10-24-28)

Stinson "Detroiter" SM-1B as 6PCLM with 220 h.p. Wright J5 engine; all ser. nos. at 3485 lbs. gross wt. Ser. #M-241, M-253, M-254 as 3PCLM of same wt. Refer to ATC #16.

#2-25 (7-6-28)

Travel Air "Smith's Incubator" as 3POLB with 120 h.p. Anzani (10 cyl.) engine; for ser. #277 at 2180 lbs. gross wt. Was formerly a model 2000 with Curtiss OX-5 engine.

#2-26 (6-14-28)

Travel Air "Huff-Daland" as 3POLB with 90 h.p. Curtiss OX-5 engine; for ser. #501-502 at 2180 lbs. gross wt. Used for crop-dusting in Louisiana.

#2-27 (2-23-29)

Travel Air 5000 as 5PCLM with 220 h.p. Wright J5 engine; for all ser. nos., but for relicensing only at 3600 lbs. gross wt. Refer to ATC #100 (Vol. 1-246) for photo.

#2-28

Travel Air 6000; this approval superseded by ATC #100.

#2-29 (12-1-28)

Thaden "Argonaut" T-1 as 7PCLM with 420 h.p. Wasp A engine; one of the earliest of all-metal airplanes. For all ser. nos. at 5320 lbs. gross wt.

#2-30 (12-21-28)

Boeing 40-B Mod. as 4POLB with 525 h.p. Hornet A engine; for ser. #882, 1095 at 6070 lbs. gross wt. Refer to ATC #2 and #27 for info on similar examples.

#2-31

Curtiss "Liberty-Falcon"; approval superseded by ATC #103.

#2-32 (2-7-29)

Ford-Stout "Tri-Motor" 5-AT-A as 14PCLM with three 420 h.p. Wasp C engines; for ser. #2-3 at 12,150 lbs. gross wt. Both airplanes delivered to Northwest Airlines.

#2-33 (1-12-29)

Brunner-Winkle "Bird" A as 3POLB with 90 h.p. Curtiss OX-5 engine; for ser. #1000-1008 at 2150 lbs. gross wt. Refer to ATC #101 (Vol. 2-10) for photo.

#2-34 (1-4-29)

Fokker F-7 as 10PCLM with 525 h.p. Hornet A engine; for ser. #617 at 7553 lbs. gross wt. This airplane delivered to Standard Air Lines.

#2-35 (2-16-29)

Travel Air W-4000 as 3POLB with 110 h.p. Warner engine; for all examples at 2276 lbs. gross wt. See ATC #112 for more details.

#2-36 (12-17-28)

Sikorsky S-38-AH as 6PCAmB with two 525 h.p. Hornet A engines; for ser. #14-6 at 10,480 lbs. gross wt. Delivered to "Liberty" magazine company. Refer to ATC #60

#2-37 (2-29-29)

Curtiss "Falcon" as 2POLB with 435 h.p. Curtiss D-12-D engine; for ser. #5 at 4658 lbs. gross wt.

#2-38 (1-19-29)

New Standard GD-24 as 5POLB with 180 h.p. Hisso E engine; for ser. #101, 102, 104 at 3400 lbs. gross wt. Refer to ATC #107 for similar airplanes.

#2-39 (10-11-29)

Avro "Avian" as 2POLB with 80-85 h.p. Cirrus Mk. 2 or Mk. 3 engine; for ser. #135-143, 147, 157, 166, 169, 186-190, 194-197, 242-281 at 1450 lbs. gross wt. Mfgd. by A.V. Roe & Co., Ltd. in England & distributed by Air Associates, Inc. in the U.S.A.

#2-40 (2-2-29)

Fairchild FC-2C as 5PCLM with 160 h.p. Curtiss C-6 engine; for ser. #57, 58, 87, 89 at 3300 lbs. gross wt. All 4 airplanes operated by Curtiss Flying Service.

#2-41 (8-2-29)

DH Moth (British) as 2POLB with 85 h.p. DH Gipsy engine; for ser. #341, 814, 885, 886, 910-913, 924-926, 978-980, 1044, 1063, 1064, 2A thru 6A at 1650 lbs. gross wt.

#2-42 (2-16-29)

Travel Air 2000 as 3POL/SB with 90 h.p. Curtiss OX-5 engine; for ser. #280 at 2351 lbs. gross wt. on Edo twin-float gear. Refer to ATC #30 for more info.

#2-43 (2-21-29)

Boeing 64 "Navy Trainer" as 2POL/SB with 220 h.p. Wright J5 engine; 2 a/c only at 2735 lbs. gross wt.

#2-44 (2-23-29)

Kreider-Reisner C-5 as 3POLB with 110 h.p. Warner engine; for ser. #179, 181, 255 at 2150 lbs. gross wt. C-3 models were eligible for C-5 modification. See ATC #97.

#2-45 (3-1-29)

Douglas M-4 Spl. as 3POLB with 525 h.p. Hornet A engine; for ser. #314. This a/c was formerly a mail-plane & converted into a blind-flight trainer. Refer to ATC #6.

#2-46 (6-28-29)

Buhl Airsedan CA-8 as 6PCLB with 450 h.p. Wasp C engine; for ser. #34 and 39 at 6100 lbs. gross wt. Refer to ATC #99 (Vol. 1-244) for photo & more details.

#2-47 (3-23-29)

Aeromarine-Klemm AKL-25A as 2POLM with 40 h.p. Salmson AD-9 engine; for ser. #1, 2, 4 at 1325 lbs. gross wt. Refer to ATC #121 (Vol. 2-61) for photo & specs.

#2-48 (3-19-29)

American Eagle A-1 Spl. as 3POLB with 128 h.p. Ryan-Siemens (Siemens-Halske SH-12) engine; for ser. #150 at 1960 gross wt. The Quick radial & Salmson 120 also tested.

#2-49 (3-22-29)

Butler "Blackhawk" as 3POLB with 220 h.p. Wright J5 engine; for ser. #103 (C-521) at 2859 lbs. gross wt. Refer to ATC #135 (Vol. 2-106) for photo. Del. to Art Goebel.

#2-50 (3-26-29)

Ryan "Brougham" B-3 as 5PCLM with 300 h.p. Wright R-975 engine; for ser. #179, 183-186 at 3704 lbs. gross wt. Marvel Crosson in cockpit.

#2-51 (5-2-29)

Buhl "Airsedan" CA-6 Spl. as 4PCLB with 300 h.p. Wright R-975 engine; for ser. #43 at 4050 lbs. gross wt. Refer to ATC #128 for similar airplanes.

#2-52 (3-28-29)

Bellanca CH-300 as 6PCLM with 300 h.p. Wright R-975 engine; for ser. #129 at 4050 lbs. gross wt. Refer to ATC #129 for similar airplanes.

#2-53 (3-29-29)

Stearman C2K as 3POLB with 128 h.p. Siemens-Halske SH-12 engine; for ser. #117 (C-4713) at 2400 lbs. gross wt. Later upgraded to C3K and then to C3B specs.

#2-54 (3-30-29)

Keystone "Pathfinder" K-47A as 11PCLB with three 220 h.p. Wright J5 engines; for ser. #137 (NC-1612) at 10,600 lbs. gross wt. Del. to West Indian Aerial Express & later used by Pan American Airways System. Airplane owned by Basil Rowe.

#2-55 (4-15-29)

American Eagle A Spl. as 3POLB with 150-180 h.p. Hispano-Suiza A or E engine; for ser. #103, 168, 171, 181, 200, 206, 230, 240, 255-259, 276, 286, 290, 291, 297, 313, 315, 392 at 2463 lbs. gross wt. Refer to ATC #17 for basic A-1 series.

#2-56 (4-1-29)

Bourdon "Kitty Hawk" B-2 as 3POLB with 113 h.p. Siemens-Halske SH-14 engine; for ser. #2 thru 7 at 1950 lbs. gross wt. Refer to ATC #134 (Vol. 2-102) for photo.

#2-57 (6-12-29)

International (Fisk) F-17-H as 3POLB with 150-180 h.p. Hispano-Suiza A or E engine; for ser. #34 thru 68 at 2706 lbs. gross wt. Refer to ATC #155 (Vol. 2-157) for photo.

#2-58 (4-12-29)

Stearman C3L as 3POLB with 130 h.p. Comet engine; for ser. #169 (NC-6438) at 2525 lbs. gross wt. Later upgraded to C3C with "Hisso" engine & then to C3B with Wright J5 engine. Registered to Walter Varney.

#2-59 (7-9-29)

Curtiss "Fledgling" as 2POLB with 6 cyl. Curtiss "Challenger" engine of 170 h.p.; as a convertible military airplane at 2701 lbs. gross wt. For ser. #1, B-1 and up when modified to conform. See ATC #191 for more details on similar a/c.

#2-60 (4-13-29)

Stinson "Detroiter" SM-1D-300 as 6PCLM with 300 h.p. Wright R-975 (J6-9-300) engine; for ser. #300 thru 307 at 4300 lbs. gross wt. Refer to ATC #74, 76, 77, 78 for SM-1D series. This approval amended 11-7-29.

#2-61 (4-19-29)

Crown "Custombilt" B-3 as 2POLB; this approval superseded by ATC #199.

#2-62 (7-31-29)

Metal Aircraft "Flamingo" G-2-W as 8PCLM with 450 h.p. Wasp SC-1 engine; for ser. #4 thru 11 at 5800 lbs. gross wt. Refer to ATC #192 (Vol. 2-268) for photo.

#2-63 (5-14-29)

Metal Aircraft "Flamingo" G-2 as 6PCLM with 450 h.p. Wasp engine; for ser. #2 (NC-588) at 5718 lbs. gross wt. See ATC #192 for similar examples.

#2-64 (6-10-29)

Boeing 40-B Mod. as 5PO/CLB with 525 h.p. Hornet A engine; converted from 40-A. For all ser. nos. at 6075 lbs. gross wt. See ATC #2 and 27 for more details on the type.

#2-65 (6-24-29)

DH Moth 60-G as 2POLB with 85 h.p. American-Gipsy engine; for ser. #1A and 1B at 1550 lbs. gross wt. Refer to ATC #197 for similar a/c with Wright-Gipsy engine.

#2-66 (5-13-29)

Lincoln-Page LP-3A as 3POLB with 150-180 h.p. Hispano-Suiza A or E engine; for ser. #247, 250 & up at 2718 lbs. gross wt. Refer to ATC #28 for similar airplanes.

#2-67 (5-13-29)

Metal Aircraft "Flamingo" G-2-H as 6PCLM with 525 h.p. Hornet A engine; for ser. #3 (NC-9304) at 5890 lbs. gross wt. Refer to ATC #192 for similar airplanes.

#2-68 (5-15-29)

Sikorsky S-38-B Spl. as 11PCAmB with two 450 h.p. Wasp engines; for ser. #114-7 (NC-9143) & 114-8 (NC-9144) at 10,480 lbs. gross wt. Ser. #114-8 later del. to U.S. Navy.

#2-69 (5-16-29)

Sikorsky S-38-B Spl. as 4PCAmB with two 450 h.p. Wasp engines; for ser. #114-9 (NC-9137) at 10,480 lbs. gross wt. Ser. #114-9 later del. to U.S. Navy.

#2-70 (5-22-29)

Stearman C3K a 3POLB with 128 h.p. Siemens-Halske SH-12 engine; for ser. #109 (C-4098) at 2400 lbs. gross wt. First as C2K and later converted to C3B.

#2-71 (5-31-29)

Curtiss "Falcon" as 2POLB with 600 h.p. Curtiss "Conqueror" engine; for ser. #6 (C-310E) at 4560 lbs. gross wt. Refer to ATC #213 for similar airplanes.

#2-72 (7-24-29)

Buhl Airsedan CA-3D Spl. as 4PCLB with 300 h.p. Wright R-975 engine; for ser. #45, 53 and up at 3200 lbs. gross wt. See ATC #163 (Vol. 2-179) for photo.

#2-73 (5-31-29)

Stinson "Junior" SM-2AA as 4PCLM with 165 h.p. Wright R-540 engine; for ser. #1046-1055 at 3152 lbs. gross wt. See ATC #145 (Vol. 2-132) for photo.

#2-74 (6-4-29)

Sikorsky S-38-B Spl. as 6PCAmB with two 450 h.p. Wasp engines; for ser. #114-14 at 10,480 lbs. gross wt. Del. as "Ancol" to Andian National Corp. in Cartagena, Colombia.

#2-75 (6-12-29)

Metal Aircraft "Flamingo" G-2-H as 8PCLM with 525 h.p. Hornet A engine; for ser. #9 and up at 6000 lbs. gross wt. Refer to ATC #192 for more details.

#2-76 (6-11-29)

Travel Air A-6000-A as 7PCLM with 450 h.p. Wasp C engine; for ser. #840, 892, 963, 981, 1078, 1084, 1097, and 1098 at 5500 lbs. gross wt. Refer to ATC #116 for similar examples of this type.

#2-77 (6-14-29)

Spartan C3-3 as 3POLB with 170 h.p. Curtiss (6 cyl.) "Challenger" engine; for ser. #101 and up at 2606 lbs. gross wt. Refer to ATC #195 for airplanes of this same basic design.

#2-78 (6-14-29)

Spartan C3-4 as 3POLB with 115 h.p. Axelson A (Floco) engine; for ser. #101 and up at 2486 lbs. gross wt. Refer to ATC #195 for airplanes of this same basic design.

#2-79 (6-14-29)

Spartan C3-5 as 3POLB with 165 h.p. Wright R-540 engine; for ser. #101 and up at 2587 lbs. gross wt. Later converted to C3-165; see ATC #195 for similar examples. Ser. #101 used as prototype for 4 different models.

#2-80 (6-14-29)

Ford "Tri-Motor" 6-AT-S Spl. as 12 PCSM with three 300 h.p. Wright R-975 engines for ser. #6-ATS-1 on Brewster floats at 12,500 lbs. gross wt. Refer to ATC #173 (Vol. 2-210) for photo as del. to RCAF.

#2-81 (6-18-29)

Kari-Keen "Coupe" as 2PCLM with 55 h.p. Velie M-5 engine; for ser. #Q-210 and up at 1400 lbs. gross wt. Refer to ATC #331 for similar airplanes.

#2-82 (6-27-29)

DH Moth (American) 60-GM; this approval superseded by ATC #197 for American-built airplanes.

#2-83 (6-21-29)

Boeing 100 Spl. as 2POLB with 450 h.p. Wasp C-1 engine; built on order for millionaire Howard Hughes. For ser. #1094 (NR-247K) only at 2694 lbs. gross wt. Refer to ATC #133 for same type as single-place version. Hughes' airplane later converted to 1POLB with NACA engine cowl & extensive fairing.

#2-84 (7-2-29)

Travel Air (Curtiss-Wright) "Speedwing" D-4000 as 1 or 3POLB with 220 h.p. Wright "Whirlwind" (J5) engine; for ser. #690 and up at 2650 lbs. gross wt. Refer to ATC #32 for the standard "4000" series. This airplane was an all-time favorite for the classic "sportsman-pilot."

#2-85 (6-24-29)

Keystone "Patrician" K-78 as 20PCLM with three 525 h.p. "Cyclone" engines; for ser. #186 at 15,000 lbs. gross wt. Refer to ATC #260 for more details.

#2-86 (6-28-29)

Fairchild 51 as 5PCLM with 300 h.p. Wright R-975 engine; mod. from FC-2 type. For ser. #157 at 4000 lbs. gross wt. Refer to ATC #357 for similar examples.

#2-87 (7-1-29)

Aeromarine-Klemm AKL-25 as 2POSM with 40 h.p. Salmson AD-9 engine; for ser. #2 and up at 1490 lbs. gross wt. on floats. Refer to ATC #121 for photo.

#2-88

Doyle "Oriole" O-2; this approval superseded by ATC #247.

#2-89 (7-3-29)

Stinson "Detroiter" SM-6B as 7PCLM with 450 h.p. Wasp engine; for ser. #2000-2001 at 5000 lbs. gross wt. Refer to ATC #217 for similar examples. "Eddie" Stinson foreground.

#2-90 (8-7-29)

Monoprep as 2POLM with 55 h.p. Velie engine; for ser. #6004-6006, 6013-6017, 6025-6035 at 1360 lbs. gross wt. Refer to ATC #218 for similar examples.

#2-91 (7-8-29)

Curtiss-Robertson "Robin" C Spl. as 3PCLM with 170 h.p. Curtiss "Challenger" engine; for ser. #180 and 210 at 2440 lbs. gross wt. See ATC #69 for similar examples.

#2-92

"Cardinal" C2-60; this approval superseded by ATC #273.

#2-93 (7-17-29)

Laird LCA-A as 6PCLB with 450 h.p. Wasp engine; for ser. #163 at 5338 lbs. gross wt. Only one built.

#2-94 (7-22-29)

Kreutzer K-2 as 6PCLM with 1 LeBlond 90 and two LeBlond 60 engines; for ser. #104 at 4445 lbs. gross wt.

#2-95 (1-15-30)

Savoia-Marchetti S-56 with 90 h.p. Kinner K5 engine; for ser. #2 and 3 at 2100 lbs. gross wt.

#2-96 (8-2-29)

Savoia-Marchetti S-56 with 90 h.p. Kinner K5 engine; for ser. #50603 at 1870 lbs. gross wt. See ATC #287.

#2-97 (7-23-29)

Ford Tri-Motor 6-AT as 16PCLM with three 300 h.p. Wright R-975 engines; for ser. #4 and up at 12,176 lbs. gross wt. Refer to ATC #173 for similar examples.

#2-98 (7-27-29)

Bach 3-CT-5 as 10PCLM with 450 h.p. Wasp in nose & two 130 h.p. Comet engines in wings; for ser. #2 at 8000 lbs. gross wt. Refer to ATC #114 for similar examples.

#2-99

Parks P-2; this approval superseded by ATC #200.

#2-100 (7-27-29)

International F-17 as 3POLB with 90 h.p. Curtiss OX-5 engine; for ser. #40-100 at 2356 lbs. gross wt.

#2-101 (7-26-29)

Waco "Ten" (220) as 3POLB with 220 h.p. Wright J5 engine; for ser. #A-117 and up at 2310 lbs. gross wt. Refer to ATC #41 for similar examples.

#2-102 (7-30-29)

Waco "Ten" (220) as 3POSB with 220 h.p. Wright J5 engine; for ser. #818 and A-116 at 2660 lbs. gross wt. on Edo floats. "Bob" Fogg operated ser. #818. See ATC #41.

#2-103 (7-31-29)

Alexander "Eaglerock" A-14 as 3POLB with 165 h.p. Wright R-540 engine; for ser. #849 and up at 2588 lbs. gross wt. Refer to ATC #139 and 141 for similar airplanes.

#2-104 (2-18-30)

Bach "Air Yacht" 3-CT-S as 9PCLM with 450 h.p. Wasp engine in nose & two 220 h.p. Wright J5 engines in wings; for ser. #8 at 8000 lbs. gross wt. Refer to ATC #172.

#2-105 (7-31-29)

Mahoney-Ryan B3A as 6PCLM with 220 h.p. Wright J5 engine; for ser. #210 (NC-311K) and up at 3700 lbs. gross wt. Refer to ATC #104 for similar airplanes.

#2-106 (8-9-29)

Rearwin "Ken-Royce" 2000-C as 3POLB with 170 h.p. Curtiss "Challenger" engine; for ser. #101 and up at 2380 lbs. gross wt. Refer to ATC #232 for similar airplanes.

#2-107 (8-12-29)

Golden Eagle "Chief" as 2POLM with LeBlond 90 (7-D) engine; for ser. #803 and up at 1480 lbs. gross wt. Refer to ATC #202 for similar airplanes.

#2-108 (8-15-29)

Atlantic Coast F5L as 14PCFbB with two 400 h.p. Liberty engines; for ser. #1 thru 5 at 13,600 lbs. gross wt. Airplanes used for air-ferry.

#2-109 (8-16-29)

"Monocoach" as 4PCLM with 220 h.p. Wright J5 engine; for ser. #5002-5007 at 3092 lbs. gross wt. Refer to ATC #201 for similar examples.

#2-110 (8-16-29)

Arrow Sport "Pursuit" as 2POLB with 90-100 h.p. Kinner K5 engine; for ser. #412-435 at 1529 lbs. gross wt. Refer to ATC #115 for same basic design.

#2-111 (8-22-29)

Aeromarine-Klemm AKL-26 as 2POSM with 65 h.p. LeBlond 5D engine; for ser. #2-32 and up at 1590 lbs. gross wt. on twin-float gear. See ATC #203 (Vol. 3-17) for photo.

#2-112 (8-23-29)

Fleet Model 3 as 2POLB with 165 h.p. Wright R-540 engine; for ser. #166 at 2000 lbs. gross wt. Refer to ATC #131 for more details on this type.

#2-113 (3-8-30)

Schreck 17-HT-4 as 4POFbB with 180 h.p. Hispano-Suiza engine; for ser. #133 and 162 at gross wts. listed in Engrg. Memo. Refer to ATC #361 (Vol. 3-195) for photo.

#2-114 (8-27-29)

Taylor "Chummy" B-2 as 2POLM with 90 h.p. Kinner K5 engine; for ser. #9 and up at 1643 lbs. gross wt.

#2-115 (8-29-29)

Moth 60-GM; this approval superseded by ATC #197.

#2-116 (9-4-29)

Keystone-Loening C2C as 9PCAmB with 525 h.p. Wright "Cyclone" engine; for ser. #216, 232, 244 at 6135 lbs. gross wt. Refer to ATC #90 for similar examples.

#2-117 (9-6-29)

Command-Aire 5C-3C as 3POLB with 165 h.p. Wright R-540 engine; for ser. #W-92 (C-932E) at 2490 lbs. gross wt. Refer to ATC #233 (Vol. 3-100) for photo.

#2-118 (9-6-29)

Avileo Liore et Olivier Le'O (French) as 2PCFbM with 180 h.p. Salmson 9AC engine; for ser. #2 and 3 at 2165 lbs. gross wt.

#2-119 (9-6-29)

Davis V-3 as 2POLM with LeBlond 60 engine; for ser. #101-123 at 1328 lbs. gross wt. See ATC #256.

#2-120 (9-6-29)

Monocoupe 113 Spl. as 2PCLM with 110 h.p. Warner engine; for ser. #321 at 1650 lbs. gross wt. Phoebe Omlie, famous aviatrix, in foreground.

#2-121 (9-6-29)

American Eagle "Phaeton"; this approval was superseded by ATC #283.

#2-122 (9-6-29)

Curtiss "Kingbird" J (formerly Kingbird C) as 6PCLM with two 240 h.p. Wright R-760 engines; for ser. #G-1 at 5600 lbs. gross wt. See ATC #347 for photo.

#2-123 (9-7-29)

American Eagle "Phaeton" R-540; this approval superseded by ATC #282.

#2-124 (9-7-29)

Stearman C3B as 3POSB with 220 h.p. Wright J5 engine; for ser. #245 on Edo P floats at 2850 lbs. gross wt. Refer to ATC #55 for landplane version.

#2-125 (9-7-29)

Hamilton Metalplane H-47 as 7PCSM with 525 h.p. Hornet A engine; for ser. #58 on twin-float gear at 6375 lbs. gross wt. Refer to ATC #94 for landplane version.

#2-126

Mercury "Chic" T-2; this approval superseded by ATC #235.

#2-127

Fairchild 42; this approval was cancelled by ATC #242.

#2-128 (9-10-29)

Monoprep as 2POLM with 55 h.p. Velie M-5 engine; for ser. #6050-6055 at 1288 lbs. gross wt. Refer to ATC #218 (Vol. 3-58) for photo.

#2-129 (2-28-29)

Hamilton H-47 Spl. as 8PCLM with 525 h.p. Wright "Cyclone" engine; all ser. nos. when mod. to conform at 5750 lbs. gross wt. Refer to ATC #94.

#2-130

Dornier "Wal" to be built by Ford-Stout under license; the approval was cancelled before any production was undertaken.

#2-131 (9-26-29)

Courier PB-1 as 3PCLM with 90 h.p. Kinner K5 engine; for ser. #100 at 2095 lbs. gross wt. Refer to ATC #285 (Vol. 3-240) for photo.

#2-132 (9-26-29)

Curt-Rob. "Robin" B Spl. as 3PCLM with 150 h.p. Hispano-Suiza A engine; for ser. #112 at 2560 lbs. gross wt. It seems that other ser. nos. were eligible when modified to conform.

#2-133 (10-4-29)

Buhl "Airsedan" CA-6W as 4PCLB with 450 h.p. Wasp engine; for ser. #48 (NC-9633) at 4200 lbs. gross wt. Refer to ATC #128 for more details.

#2-134 (10-4-29)

Monosport 1 as 2PCLM with 110 h.p. Warner engine; for ser. #2000-2002, 2005 at 1650 lbs. gross wt. Refer to ATC #249 (Vol. 3-144) for photo.

#2-135 (10-4-29)

Monosport 2 as 2PCLM with 100 h.p. Kinner K5 engine; for ser. #2003, 2004 at 1650 lbs. gross wt. Refer to ATC #250 (Vol. 3-147) for photo.

#2-136 (10-4-29)

Stinson "Junior" SM-2K as 3PCLM with 100 h.p. Kinner K5 engine; for ser. #1029, 1033, 1037 at 2500 lbs. gross wt. See ATC #48 for SM-2 details.

#2-137 (10-4-29)

Command-Aire 3C3-A as 3POSB with 110 h.p. Warner engine; for ser. #W-79 (NC-916E) at 2305 lbs. gross wt. on Edo Deluxe floats. Refer to ATC #118 for landplane version.

#2-138 (10-5-29)

Travel Air (Curtiss-Wright) S-6000-B Spl. as 7PCLM with 300 h.p. Wright R-975 (J6-9-300) engine; for all ser. nos. at 4230 lbs. gross wt. These airplanes were fitted with extra seat. Refer to ATC #130 for similar examples.

#2-139 (10-9-29)

Boeing Trainer 203; this approval was cancelled because all airplanes were converted to model 203-A as per specifications of ATC #211.

#2-140 (10-9-29)

Corman 3000 "Tri-Motor" as 7PCLM with three 220 h.p. Wright J5 engines; for ser. #2 at 7652 lbs. gross wt. Refer to ATC #335 (Vol. 4-124) for photo. This was the prototype for Stinson "Tri-Motor" SM-6000.

#2-141 (10-21-29)

Thunderbird W-14 as 3POLB with 90 h.p. Curtiss OX-5 engine; for ser. #28-4 (4th airplane of 1928) at 2361 lbs. gross wt. Typical of most other OX-5 powered biplanes of this period.

#2-142 (10-23-29)

Stinson "Detroiter" SM-1D Spl. as 4PCLM with 220 h.p. Wright J5 engine; for all ser. nos. at 4500 lbs. gross wt. See ATC #74, 76, 77, 78 for similar examples.

#2-143 (10-23-29)

Stinson "Junior" SM-2AC as 4PCSM with 225 h.p. Wright R-760 (J6-7-225) engine; for ser. #1093 (NC-452H) on Edo P-2 Mod. twin-float gear at 3522 lbs. gross wt. Refer to ATC #194 for landplane version.

#2-144 (10-30-29)

Monarch A as 3POLB with 90 h.p. Curtiss OX-5 engine; for ser. #6 and up at 2150 lbs. gross wt. Typical of most other OX-5 powered biplanes of this period.

#2-145 (11-1-29)

Coffman "Ranger" A as 3PCLM with 90 h.p. Curtiss OX-5 engine; for ser. #3 & 100 thru 110 at 2132 lbs. gross wt. Designed by "Sam" Coffman.

#2-146 (11-1-29)

Monocoupe 113 Spl. as 2PCLM with 80 h.p. A-S (Armstrong-Siddeley) "Genet" engine; for ser. #317 at 1362 lbs. gross wt. "Genet" was a British-built engine. Refer to ATC #113 for more details on this series of airplanes.

#2-147 (11-2-29)

Boeing 81-B Trainer as 2POLB with 115 h.p. Axelson A engine; for ser. #1037 at 2230 lbs. gross.

#2-148 (11-15-29)

Davis D-1-K as 2POLM with 100 h.p. Kinner K5 engine; for ser. #501, 701 at 1476 lbs. gross wt.

#2-149 (11-14-29)

Stearman M-2 Spl. as 2POLB with 525 h.p. Hornet engine; for ser. #1007 at 5558 lbs. gross wt.

#2-150 (11-14-29)

DH Moth 60X as 2POLB with 76 h.p. Cirrus Mk. 2 engine; ser. #547, 626 at 1402 lbs. gross wt.

#2-151 (11-15-29)

Sierra BLW-2 as 3PCLM with 130-165 h.p. Comet engine; for ser. #2 and up at 2131 lbs. gross wt.

#2-152 (11-15-29)

Fleet 5 as 2POLB with 90 h.p. Brownback C-400 engine; for ser. #206 and up at 1620 lbs. gross wt. Refer to ATC #131 for more details. Later for Model 12 also.

#2-153 (11-18-29)

Travel Air 6000-B Spl. as 5PCLM with 300 h.p. Wright R-975 engine; for ser. #6B-2028 at 4230 lbs. gross wt. Refer to ATC #130 for similar airplanes.

#2-154 (11-20-29)

Travel Air SBC-4000 as 3POSB with 170-185 h.p. Curtiss "Challenger" engine; for ser. #1041 on Edo P floats at 2900 lbs. gross wt. See ATC #189 for landplane version.

#2-155 ((11-22-29)

Stearman 4-C as 3POLB with 300 h.p. Wright R-975 engine; for ser. #4001-4003 at 3794 lbs. gross wt. See ATC #304 (Vol. 4-19) for photo.

#2-156

Travel Air E-4000; this approval superseded by ATC #188.

#2-157 (11-26-29)

Nicholas-Beazley NB-3 as 2POLM with LeBlond 60 engine; for ser. #18 (C-9316) at 1296 lbs. gross wt.

#2-158 (12-2-29)

Bolte LW-2 as 2POLM with 90 h.p. Kinner K5 engine; for ser. #4 and up at 1525 lbs. gross wt.

#2-159 (12-4-29)

Stearman C3B Spl. as 3POLB with 220 h.p. Wright J5 engine; for all ser. nos. at 2830 lbs. gross wt. These airplanes were originally as C3B that were modified into C3MB mail-carriers & then converted back to the C3B configuration.

#2-160 (12-4-29)

Travel Air 4-P as 3POLB with 140 h.p. A.C.E. LA-1 engine; for ser. #1332 and up at 2388 lbs. gross wt.

#2-161 (12-5-29)

Bird-Wing "Imperial" as 3POLB with 165 h.p. Wright R-540 engine; for ser. #100 and up at 2270 lbs. gross wt. Designed by R.T. "Bob" McCrum, designer of original "American Eagle" biplane.

#2-162

This approval was actually allotted, but was cancelled and not used. Identification of airplane not know.

#2-163 (12-6-29)

Fokker "Amphibian" F-11-A as 6PCAmM with 525 h.p. Wright "Cyclone" engine; for ser. #901 (NC-7887) at 6000 lbs. gross wt. First had 450 h.p. Wasp engine for test. Refer to ATC #222 (Vol. 3-69) for photo.

#2-164 (12-16-29)

Paramount "Cabinaire" model 110 as 4PCLB with 110 h.p. Warner "Scarab" engine; for ser. #6 only at 2255 lbs. gross wt. Refer to ATC #265 (Vol. 3-189) for photo.

#2-165 (12-16-29)

Paramount "Cabinaire" model 110 as 3PCLB with 110 h.p. Warner "Scarab" engine; for ser. #1 thru 5 at 2252 lbs. gross wt. Refer to ATC #265 for details on this basic design.

#2-166 (12-16-29)

American Eagle "Hisso-Phaeton" as 3POLB with 150-180 h.p. Hispano-Suiza A or E (Hisso) engine; for ser. #600 and up at 2849 lbs. gross wt. Refer to ATC #283 (Vol. 3-237) for photo.

#2-167 (1-3-30)

Travel Air (Curtiss-Wright) 6000-B as 6PCLM with 300 h.p. Wright R-975 (J6-9-300) engine; for ser. #2025 at 4230 lbs. gross wt. See ATC #130 for similar airplanes.

#2-168 (1-4-30)

Viking (Schreck) "Flying Boat" 17-HMT-2 as 2POFbB with 180 h.p. (French) Hispano-Suiza E engine; for ser. #64 at 3144 lbs. gross wt. Refer to ATC #361 for more details on this series.

#2-169 (1-7-30)

Travel Air 4-D Special as 3POLB with 240 h.p. Wright R-760 engine; for ser. #1160 (NC-9961) at 2880 lbs. gross wt. See ATC #254 for similar examples.

#2-170 (1-7-30)

Sikorsky S-27-2 "Guardian" as 10PCLB with two 525 h.p. Hornet A engines; for ser. #2 at 14,500 lbs. gross wt. Tested by USAAC as bomber, but not accepted.

#2-171 (2-3-30)

Emsco B-2 Tri-Motor as 8PCLM with three 170 h.p. Curtiss "Challenger" engines; for ser. #2-6 at 7233 lbs. gross wt. See ATC #400 (Vol. 4-313) for photo.

#2-172 (1-9-30)

Fokker F-11-AHB as 8PCAmM with 575 h.p. Hornet B engine; for ser. #4 at 7200 lbs. gross wt. Built on order for "Gar" Wood.

#2-173 (1-11-30)

Miami Maid MM-201 as 4PCFbM with 300 h.p. Wright R-975 engine; for ser. #52 at 3750 lbs. gross wt.

#2-174 (1-16-30)

Stinson SM-1 Spl. as 6PCLM with 220 h.p. Wright J5 engine; for ser. #M-219, M-224, M-234, M-249, M-250-251, M-255 at 3485 lbs. gross wt.

#2-175 (1-17-30)

Bach 3-CT-9 Spl. as 7PCLM with 450 h.p. Wasp in nose & two Wright R-760 engines in wings; for ser. #19 at 8000 lbs. gross wt. See ATC #271 (Vol. 3-204) for photo.

#2-176 (1-31-30)

Whittelsey "Avian" Mk. 4M as 2POLB with 85 h.p. ADC Cirrus engine; for ser. #101 and up at 1579 lbs. gross wt. This is American-built version of British "Avian."

#2-177 (2-5-30)

Stearman 4CM as 1POLB with 300 h.p. Wright R-975 engine; for ser. #4012 at 3800 lbs. gross wt.

#2-178 (2-5-30)

Travel Air D4D as 3POLB with 240 h.p. Wright R-760 engine; for ser. #1340, 1372, 1374, 1376, 1391 at 2650 lbs. gross wt. H. Lloyd Child, test-pilot, in foreground.

#2-179 (2-10-30)

Bach "Air Yacht" 3-CT-98; this approval was superseded by ATC #299.

#2-180 (2-12-30)

Driggs "Skylark" model 3; this approval was superseded by ATC #303.

#2-181 (2-18-30)

Alexander "Bullet" C-7; this approval was superseded by ATC #318.

#2-182 (2-21-30)

Watkins "Skylark" model SL as 2POLM with 65 h.p. LeBlond 5D engine; for ser. #101 and up at 1477 lbs. gross wt. The prototype had 55 h.p. Velie M-5 engine. Designed by "Chet" Cummings.

#2-183 (2-30)

Curtiss "Kingbird" J-2 as 8PCLM with two Wright R-760 engines of 240 h.p. each; for ser. #G-3 at 6100 lbs. gross wt. See ATC #347 (Vol. 4-156) for photo.

#2-184 (2-27-30)

Fokker "Super Universal" as 'PCLM with 450 h.p. Wasp C-1 engine; for ser. #841 at 5550 lbs. gross wt. Refer to ATC #52 for similar examples.

#2-185 (2-28-30)

Bellanca CH-300 "Pacemaker Special" as 6PCLM with 300 h.p. Wright R-975 (J6-9-300) engine; for ser. #182 at 4300 lbs. gross wt. Ser. #182 was NC-859N. Refer to ATC #129 for similar examples.

#2-186 (2-28-30)

Swallow R-760 as 3POLB with 225 h.p. Wright R-760 (J6-7-225) engine; for ser. #1042 at 2700 lbs. gross wt. See ATC #51 for similar examples.

#2-187 (3-4-30)

Keystone-Loening C2H as 9PCAmB with 525 h.p. Hornet A engine; for ser. #243, 244, 245 at 6135 lbs. gross wt. Refer to ATC #91 for similar examples.

#2-188 (3-4-30)

St. Louis "Cardinal" C2-110 Spl. as 2PCLM with 100 h.p. Kinner K5 engine; for ser. #118 at 1575 lbs. gross wt. See ATC #277 (Vol. 3-219) for photo.

#2-189 (3-4-30)

Laird LCB-300 (LC-1B300) as 3POLB with 300 h.p. Wright R-975 engine; for ser. #176, 184, 186 at 3022 lbs. gross wt. See ATC #353 (Vol. 4-174) for photo.

#2-190

Sikorsky S-38-BH; this approval superseded by ATC #356.

#2-191 (3-7-30)

Star "Cavalier" D as 2PCLM with 80 h.p. "Genet" engine; for ser. #116, 120 at 1400 lbs. gross wt. Refer to ATC #255 for similar airplanes.

#2-192 (3-13-30)

Curtiss-Robertson "Robin" C-1 Spl. as 3PCLM with 185 h.p. Curtiss "Challenger" engine; for ser. #668 with Goodyear airwheels & rigid landing gear at 2600 lbs. gross wt. Refer to ATC #143 for more details.

#2-193 (3-15-30)

Pitcairn "Fleetwing" PA-4W as 3POLB with 110 h.p. Warner engine; for ser. #1 at 1920 lbs. gross wt.

#2-194 (3-18-30)

Granville "Gee Bee" A as 2POLB with 100 h.p. Kinner K5 engine; for ser. #P-1 thru P-8 at 1654 lbs. gross wt. See ATC #398 for Granville Bros. background.

#2-195 (3-24-30)

Davis D1-L as 2POLM with 90 h.p. Lambert R-266 engine; for ser. #301 at 1377 lbs. gross wt. Refer to ATC #317 for more details.

#2-196 (3-26-30)

Curtiss "Kingbird" J-3 as 2-6PCLM with two 300 h.p. Wright R-975 engines; for ser. #G-2 (NC-310N) at 5339 lbs. gross wt. Refer to ATC #347 for "Kingbird" series.

#2-197 (3-26-30)

Ogden "Osprey" PC as 6PCLM with three 90 h.p. A.C.E. Cirrus engines; for ser. #102, 103 at 4500 lbs. gross wt. see ATC #332 (Vol. 4-115) for photo.

#2-198 (4-3-30)

Curtiss "Robin" 4C1 as 3PCLM with 170 h.p. Curtiss "Challenger" engine; for ser. #700, 767, 769 at 2600 lbs. gross wt. Refer to ATC #270 for similar examples.

#2-199 (4-5-30)

Ireland "Neptune" N2B Spl. as 5POFbB with 300 h.p. Wright R-975 engine; for ser. #32 (NC-106H) at 4256 lbs. gross wt. Refer to ATC #153 for N2B series.

#2-200 (4-5-30)

Fokker "Amphibian" F-11-AHB as 10PCAmM with 575 h.p. Hornet B engine; for ser. #906 (NC-339N) at 7200 lbs. gross wt. Used as air-ferry.

#2-201 (4-7-30)

Command-Aire 3C3 as 3POLB with 90 h.p. Curtiss OX-5 engine; for ser. #640 thru 656 at 2118 lbs. gross wt. Refer to ATC #53 for earlier examples.

#2-202 (4-12-30)

Timm "Collegiate" C-185 as 2POLM with 185 h.p. Curtiss "Challenger" engine; for ser. #103 at 2212 lbs. gross wt. First as model C-170. Refer to ATC #180.

#2-203 (4-16-30)

Fairchild 42 Spl. as 5PCLM with 300 h.p. Wright R-975 engine; for ser. #3 (NC-81M) at 4300 lbs. gross wt. See ATC #242 for more details.

#2-204 (4-16-30)

Command-Aire BS-14 trainer as 2POLB with 110 h.p. Warner engine; for ser. #W-138 only at 1983 lbs. gross wt. Refer to ATC #151 for similar airplanes.

#2-205 (4-30-30)

Bellanca "Skyrocket" CH-400 Spl. as 5PCLM with 450 h.p. Wasp SC-1 engine; for ser. #601 (NC-179N) at 4600 lbs. gross wt. See ATC #319 for "Skyrocket" series.

#2-206 (5-1-30)

Savoia-Marchetti S-55-P as 15PCFbM with two 500 h.p. Issotta-Fraschini "Asso" engines; for ser. #10514, 10517, 10520 at 16,100 lbs. gross wt. See ATC #287 (Vol. 3-248) for photo. NC-105H was with Airvia Transport.

#2-207 (5-1-30)

Avro "Avian" Spl. as 2POLB with 90 h.p. Wright "Gipsy" engine; for ser. #136 at 1450 lbs. gross wt. Del. to Chas. Lawrance, father of famous Wright "Whirlwind" engine.

#2-208 (5-2-30)

Fairchild KR-32 as 2POLB with 165 h.p. Wright R-540 engine; for ser. #385 at 2210 lbs. gross wt. Refer to ATC #162 for same basic design.

#2-209 (5-8-30)

Timm "Collegiate" C-165 as 2POLM with 165 h.p. Comet 7-E engine; for ser. #101, 102 at 2176 lbs. gross wt. Refer to ATC #180 for more details.

#2-210 (5-9-30)

Curtiss-Robertson "Thrush" J Spl. as 6PCLM with 240 h.p. Wright R-760 engine; for ser. #1005 (C-553N) at 3800 lbs. gross wt. See ATC #261 for similar examples.

#2-211 (5-9-30)

Hodkinson HT-1 "Tri-Motor" as 7PCLB with three 185 h.p. Curtiss "Challenger" engines; for ser. #101 at 6780 lbs. gross wt. Reportedly shipped to So. America.

#2-212 (5-10-30)

Zenith Z6A as 7PCLB with 450 h.p. Wasp SC-1 engine; for ser. #3 at 4394 lbs. gross wt.

#2-213 (5-16-30)

Bellanca "Skyrocket" CH-400 Spl. as 6PCLM with 450 h.p. Wasp engine; for ser. #606 (NC-548V) at 4600 lbs. gross wt. Refer to ATC #319 for similar examples.

#2-214 (5-14-30)

Overcashier O-12 as 3PCLM with 90 h.p. Curtiss OX-5 engine; for ser. #1-4 at 2344 lbs. gross wt.

#2-215 (5-19-30)

Boeing Flying Boat B1E as 5PCFbB with 450 h.p. Wasp engine; for ser. #1074 at 4500 lbs. gross wt. Refer to ATC #64 for similar examples.

#2-216 (5-19-30)

Aeronca C-2; approval superseded by ATC #351.

#2-217 (5-28-30)

Lincoln-Page LP-3 "Axelson Spl." as 3POLB with 115 h.p. Axelson A engine; for ser. #268 at 2529 lbs. gross wt.

#2-218 (6-5-30)

Atlanta "Tri-Motor" PW-1 as 8PCLM with three 240 h.p. Wright R-760 engines; for ser. #2 at 7345 lbs. gross wt. A Geo. H Prudden design.

#2-219 (6-6-30)

Fleetster 17-2 Spl. as 6PCLM with 525 h.p. Hornet A engine; for ser. #3 at 5900 lbs. gross wt. Refer to ATC #291 (Vol. 3-260) for photo. Del. to F. Trubee Davison.

#2-220 (6-13-30)

Ryan "Brougham" B5A as 6PCLM with 300 h.p. Wasp Jr. engine; for ser. #200 (NC-15H) at 4000 lbs. gross wt. Ship originally had Wright R-975 engine as B5.

#2-221 (6-13-30)

Cessna DC-6B as 1-4PCLM with 240 h.p. Wright R-760 engine; for ser. #200-202, 211, 213, 216-219, 222 at 3100 lbs. gross wt. for delivery of newspapers.

#2-222 (6-13-30)

Curtiss-Wright Travel Air "Speedwing" D-4000 as 3POLB with 220 h.p. Wright J5 engine; for ser. #1391 at 2650 lbs. gross wt. Later as D4D Special.

#2-223 (6-13-30)

Ryan "Brougham" B-7 as 6PCLM with 450 h.p. Wasp engine; for ser. #255 at 4300 lbs. gross wt. Refer to ATC #262 (Vol. 3-180) for photo.

#2-224 (6-13-30)

Stinson SM-1B Spl. as 5PCLM with 220 h.p. Wright J5 engine; for ser. #M-212 at 3597 lbs. gross wt.

#2-225 (6-13-30)

Buhl CA-6B as 6PCLB with 450 h.p. Wasp engine; for ser. #53, 61-63 at 4665 lbs. gross wt.

#2-226 (6-16-30)

Bellanca CH-400 as 5PCLM with 450 h.p. Wasp C engine for ser. #605 (NC-547V) at 4530 lbs. gross wt.

#2-227 (6-17-30)

Bourdon "Kitty Hawk" B-4 as 3POLB with 100 h.p. Kinner K5 engine; for ser. #12 (NC-293M) at 1988 lbs. gross wt. See ATC #166 for similar airplanes.

#2-228 (6-20-30)

Stinson "Detroiter" SM-1DX as 3PCLM with 225 h.p. Packard Diesel engine; for ser. #303 at 4245 lbs. gross wt. Refer to ATC #74 for more data.

#2-229 (6-23-30)

Saul "Triad" 1000 as 4PCLM with three 65 h.p. LeBlond 5D engines; for ser. #1001 at 3240 lbs. gross wt. An unusual light "tri-motor."

#2-230 (6-26-30)

Pitcairn "Fleetwing" PA-4K as 3POLB with 100 h.p. Kinner K5 engine; for ser. #1 at 1920 lbs. gross wt. This a/c first had OX-5 engine.

#2-231 (7-2-30)

Fleetster 20-1 as 1-4PO/CLM with 575 h.p. Hornet B engine; for ser. #1-2 at 5900 lbs. gross wt. See ATC #320 (Vol. 4-74) for photo.

#2-232 (7-3-30)

General "Aristocrat" 102-F as 3PCLM with 165 h.p. Continental A-70 engine; for ser. #22, 27, 28, 29, 30 at 2305 lbs. gross wt. Refer to ATC #117 and 210.

#2-233 (7-9-30)

Paramount "Cabinaire" A-70 as 4PCLB with 165 h.p. Continental A-70 engine; for ser. #9 at 2600 lbs. gross wt. Refer to ATC #265 for more details.

#2-234 (7-9-30)

Travel Air 6000-B Spl. as 6PCLM with 300 h.p. Wright R-975 engine; for ser. #6B-2012 at 4230 lbs. gross wt. Refer to ATC #130 for similar examples.

#2-235 (7-12-30)

Fleet Model 2; approval superseded by ATC #131.

#2-236 (7-17-30)

Breese R-6 as 1POLM with 300 h.p. Wright R-975 engine; for ser. #9 at 3600 lbs. gross wt.

#2-237 (7-17-30)

Cessna AF Spl. as 3PCLM with 115 h.p. Axelson A engine; for ser. #141 at 2262 lbs. gross wt.

#2-238 (7-22-30)

Bellanca "Pacemaker" CH-200 Spl. as 6PCLM with 225 h.p. Packard Diesel DR-980 engine; for ser. #185, 195 at 4242 lbs. gross wt. See ATC #47 for CH-200 series.

#2-239 (7-25-30)

Timm "Collegiate" M-150 as 2POLM with 150 h.p. Maclatchie "Panther" engine; for ser. #105 at 2199 lbs. gross wt. Set endurance record of 97 T.O. & landings in 378 hrs. without shutting down engine.

#2-240 (7-26-30)

Ford 5-AT-C Spl. as 15PCLM with three 420 h.p. Wasp engines; for ser. #3 at 13,245 lbs. gross wt.

#2-241

Fairchild 51-A; this approval superseded by ATC #358.

#2-242 (7-28-30)

Ford 5-AT-C Spl. as 15PCLM with three 420 h.p. Wasp engines; for ser. #70 at 13,500 lbs. gross wt. Refer to ATC #165 for more details.

#2-243 (7-28-30)

Travel Air J4-4000 as 3POLB with 200 h.p. Wright J4 engine; for ser. #1339 at 2654 lbs. gross wt.

#2-244 (7-29-30)

Fairchild 71 as 7PCLM with 420 h.p. Wasp engine; for ser. #623 at 5500 lbs. gross wt. Refer to ATC #89.

#2-245 (7-30-30)

Waco DSO-150 as 3POSB with 150 h.p. Hispano-Suiza A engine; for ser. #D-3190 at 2404 lbs. gross wt. as landplane & 2660 lbs. as seaplane on Edo floats. See ATC #42.

#2-246 (8-4-30)

Franklin Model A as 2POLB with 65 h.p. Velie M-5 engine; for ser. #3 and up at 1312 lbs. gross wt. Refer to ATC #430 for more details.

#2-247 (8-6-30)

Pittsburgh-Thaden T-4 as 4PCLM with 300 h.p. Wright R-975 engine; for ser. #1 & 4 at 3800 lbs. gross wt. Designed by Herbert V. Thaden.

#2-248 (8-6-30)

Boeing "Tri-Motor" 80-B1 as 16PCLB with three 575 h.p. Hornet B engines; for ser. #1092 at 17,500 lbs. gross wt. Refer to ATC #206 for more details.

#2-249 (8-7-30)

Command-Aire 5C3-B as 3POLB with 115 h.p. Axelson A engine; for ser. #W-142 at 2510 lbs. gross wt. See ATC #214 for similar airplanes.

#2-250

Fairchild KR34-D; this approval superseded by ATC #376.

#2-251 (8-12-30)

Command-Aire 5C3 as 3POLB with 185 h.p. Curtiss "Challenger" engine; for ser. #W-143 at 2419 lbs. gross wt. Refer to ATC #184 for similar examples.

#2-252 (8-8-30)

Solar MS-1 as 8PCLB with 420 h.p. Wasp C engine; for ser. #101 at 5650 lbs. gross wt. Solar quit a/c mfgr. to specialize in stainless-steel exhaust systems.

#2-253 (8-9-30)

Fleet 2X Spl. as 2POLB with 100 h.p. Kinner K5 engine; for ser. #52 at 1675 lbs. gross wt. Refer to ATC #131 for similar airplanes.

#2-254 (8-12-30)

States B-2 as 2POLM with 75 h.p. Michigan-Rover engine; for ser. #102 if bag. compt. sealed, & for #101 also if changed to conform. Eligible at 1558 lbs. gross wt. Refer to ATC #349 (Vol. 4-162) for photo.

#2-255 (8-12-30)

Waco RNF Spl. as 2POLB with 110 h.p. Warner engine; for ser. #3316 at 1900 lbs. gross wt. See ATC #311 for similar examples.

#2-256 (8-13-30)

Lockheed "Vega" 5 Spl. as 5PCL/SM with 450 h.p. Wasp engine; for ser. #102 at 4265 lbs. as landplane & 4698 lbs. as seaplane on APC twin-float gear. Del to Detroit News.

#2-257 (8-13-30)

Fairchild KR-34-C as 3POSB with 165 h.p. Wright R-540 engine; for ser. #1, 900-908 at 2640 lbs. gross wt. Refer to ATC #162 for landplane version.

#2-258 (8-14-30)

Aircraft Builders "Student Prince" X as 2POLB with 90 h.p. A.C.E. Cirrus Mk. 3 engine; for ser. #101 thru 103 at 1500 lbs. gross wt.

#2-259 (8-14-30)

Fairchild 51 as 5PCSM with 300 h.p. Wright R-975 engine; for ser. #102 at 4000 lbs. gross wt. on P-4 floats. See ATC #357 for typical photo.

#2-260 (8-15-30)

Nicholas-Beazley NB-4L as 3POLM with 90 h.p. Lambert R-266 engine; for ser. #103 and up at 1511 lbs. gross wt. Refer to ATC #385 for similar examples.

#2-261

Great Lakes 2T-1E; this approval superseded by ATC #354.

#2-262 (8-20-30)

Waco BSO (BS-165) as 3POSB with 165 h.p. Wright R-540 engine; for ser. #3002 at 2738 lbs. gross wt. on Edo M twin-float gear. See ATC #168 for landplane version.

#2-263 (8-20-30)

Detroit-Ryan "Foursome" C-2 as 4PCLM with 225 h.p. Packard Diesel DR-980 engine; for ser. #401 only at 3450 lbs. gross wt. See ATC #346 for "Foursome" details.

#2-264 (8-27-30)

Nicholas-Beazley NB-4W as 3POLM with 90 h.p. Warner "Scarab Jr." engine; for ser. #105 and up at 1543 lbs. gross wt. See ATC #386 for similar examples.

#2-265 (9-2-30)

Timm "Collegiate" TC-165 as 2POLM with 165 h.p. Continental A-70 engine; for ser. #104 and up at 2230 lbs. gross wt. Refer to ATC #180 for more details.

#2-266 (9-5-30)

Mooney "Low-Wing" A-1 as 4PCLM with 100 h.p. Kinner K5 engine; for ser. #1 and up at 2250 lbs. gross wt. Note similarity to Alexander "Bullet," also designed by Al W. Mooney.

#2-267 (9-6-30)

Stinson "Junior" SM-2AB as 4PCSM with 220 h.p. Wright J5 engine; for ser. #1062, 1094 at 3450 lbs. gross wt. on Edo twin float gear. See ATC #161 for landplane version.

#2-268 (9-10-30)

Hamilton (Boeing) Metalplane H-22 as 6PCSM with 420 h.p. Wasp engine; for ser. #44 at 5725 lbs. gross wt. on Hamilton twin-float gear.

#2-269 (9-15-30)

Zenith Z-6-A as 7PCLB with 420 h.p. Wasp C engine; for ser. #3 and up at 4393 lbs. gross wt.

#2-270 (9-17-30)

Inland "Sport" S-300-DF as 2POLM with 85 h.p. LeBlond 5DF engine; for ser. #S-316-DF at 1314 lbs. gross wt. Refer to ATC #259 for more details.

#2-271 (9-18-30)

California "Cub" D-1 as 2POLM with 90 h.p. Lambert R-266 engine; for ser. #2 at 1489 lbs. gross wt. A "Larry" Brown design.

#2-272 (9-24-30)

New Standard D-29-S as 2POLB with 100 h.p. Kinner K5 engine; for ser. #1020 at 1839 lbs. gross wt. This a/c was formerly a D-29-A; see ATC #216.

#2-273 (9-25-30)

Consolidated "Fleetster" 17-2C as 7PCLM with 525 h.p. Wright "Cyclone" engine; for ser. #6 (NC-750V) at 5600 lbs. gross wt. See ATC #369 (Vol. 4-217) for photo. Later operated by Pacific-International Airways in Alaska.

#2-274 (9-29-30)

Lockheed "Vega" 5-B Mod. as 5-7PCLM with 420 h.p. Wasp engine; for ser. #101 at 4265 lbs. gross wt. Refer to ATC #227 for similar examples.

#2-275 (9-30-30)

Sikorsky S-36 as 8PCAmB or 8PCFbB with two 220 h.p. Wright J5 engines; for ser. #S-36-5 at 6800 lbs. gross wt. Used by Pan American Airways in 1927.

#2-276 (10-3-30)

New Standard (Barnard) D-31 as 2POLB with 125 h.p. Kinner B5 engine; for ser. #1006, 1025, 1033 at 1850 lbs. gross wt. Formerly as D-29-A. See ATC #324.

#2-277 (10-3-30)

Savoia-Marchetti S-55 as 11PCFbM with two 518 h.p. "Asso" engines; for ser. #1 at 15,000 lbs. gross wt. Later had two 575 h.p. Wright "Cyclone" engines.

#2-278 (10-6-30)

Stearman 4E Spl. as 3POLB with 420 h.p. Wasp engine; for ser. #4021-4022 at 3936 lbs. gross wt. Refer to ATC #292 for similar examples.

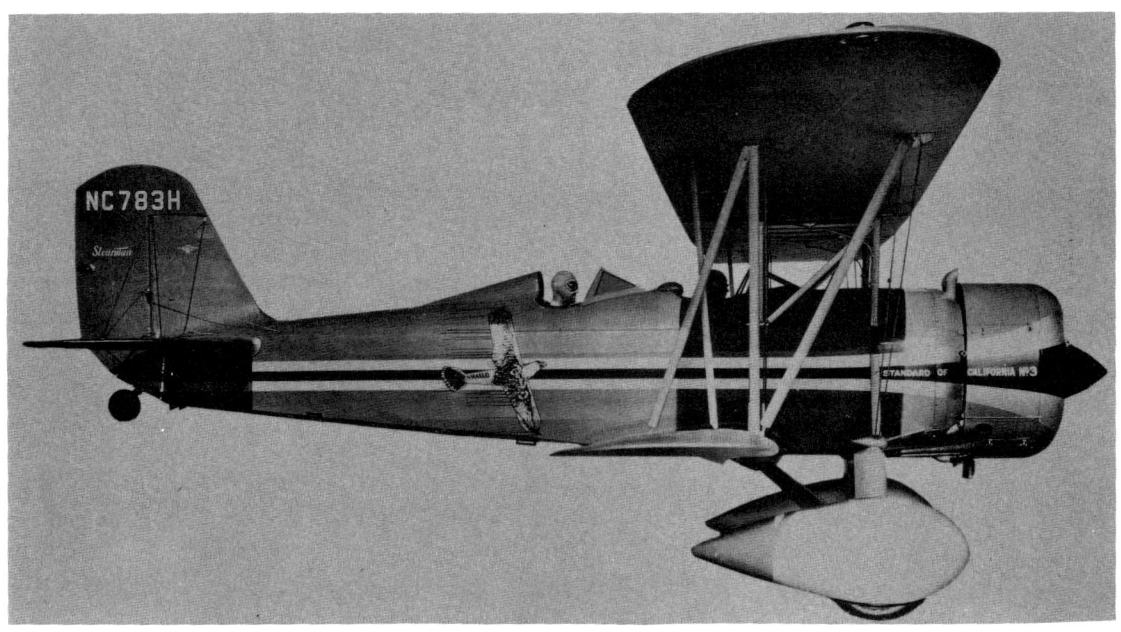

#2-279 (10-6-30)

Stearman 4EX as 2POLB with 450 h.p. Wasp SC-1 (supercharged) engine; for ser. #4020 at 3936 lbs. gross wt. This airplane del. to Standard Oil of California.

#2-280 (10-7-30)

Nicholas-Beazley NB-4G as 3POLM with 80 h.p. Armstrong-Siddeley "Genet" Mk. 2 engine; for ser. #103 and up at 1529 lbs. gross wt.

#2-281 (10-10-30)

Driggs "Skylark" 3-95 as 2POLB with 95 h.p. A.C.E. Cirrus Hi-Drive engine; for ser. #3015 and 3017 at 1535 lbs. gross wt. See ATC #303 for "Skylark" series.

#2-282 (10-10-30)

Ford Tri-Motor 5-AT-A as 13PCLM with three 420 h.p. Wasp A engines; for ser. #1 at 12,000 lbs. gross wt. This was the first 5-AT & engines were ser. #X-1, X-2, X-3. Del. to Col. C. W. Deeds of Pratt & Whitney & flown regularly from the back-lot of his estate. Refer to ATC #156.

#2-283 (10-10-30)

Buhl "Airster" CA-1WA as 3POLM with 420 h.p. Wasp engine; for ser. #64 at 3580 lbs. gross wt. This was the "sportsman's" version.

#2-284 (10-10-30)

Lockheed "Vega" 5-A Spl. as 5PCLM with 450 h.p. Wasp SC engine; for ser. #80 at 4217 lbs. gross wt. Del. to Sen. W. G. McAdoo.

#2-285 (10-13-30)

General-Western "Phantom" P-2 as 2POLM with 100 h.p. Kinner K5 engine; for ser. #101 at 1650 lbs. gross wt. Refer to ATC #482 for more details.

#2-286 (10-14-30)

Sikorsky S-41 as 13PCAmM with two 575 h.p. Hornet B engines; for ser. #1100-X at 11,504 lbs. gross wt. See ATC #418 for photo & more details.

#2-287 (10-16-30)

Curtiss "Falcon" Mailplane as 1POLB with 575 h.p. Wright "Cyclone" engine; for ser. #3 at 4650 lbs. gross wt. Del. to Pan Am-Grace in So. America.

#2-288 (10-17-30)

Fokker Tri-Motor F-10-A as 10PCLM with three 420 h.p. Wasp engines; for ser. #1036 at 13,600 lbs. gross wt. Refer to ATC #96 for similar examples.

#2-289 (10-17-30)

Overland "Sport" as 2POLB with 70 h.p. LeBlond 5DE engine; for ser. #114 and up at 1301 lbs. gross wt. Refer to ATC #417 (Vol. 5-50) for photo.

#2-290 (10-18-30)

Fairchild 71-A as 5PCLM with 420 h.p. Wasp C engine; for ser. #3501 at 5500 lbs. gross wt. Refer to ATC #289 for more details.

#2-291 (10-21-30)

Towle "Amphibian" TA-3 as 6PCAmM with two 225 h.p. Packard Diesel DR-980 engine; for ser. #1 at 5805 lbs. gross wt. Designed by "Tom" Towle.

#2-292 (10-22-30)

Fairchild KR-34-C Spl. as 1POLB with 165 h.p. Wright R-540 engine; for ser. #333 at 2368 lbs. gross wt. Used by Gulf Coast Airways as mailplane.

#2-293 (10-23-30)

Bellanca CH-400 Spl. as 6PCLM with 420 h.p. Wasp engine; for ser. #609 at 4600 lbs. gross wt.

#2-294 (10-29-30)

Travel Air S-6000-B Spl. as 4PCLM with 300 h.p. Wright R-975 engine; for ser. #964, 986, 1029, at 4230 lbs. gross wt. Refer to ATC #130 for similar examples.

#2-295 (10-30-30)

Ogden "Osprey-Pirate" as 6PCLM with three 95 h.p. Menasco B-4 engines; for ser. #105 and up at 4500 lbs. gross wt. Refer to ATC #332 for more details.

#2-296 (10-31-30)

Travel Air 6000-B Spl. as 5PCLM with 300 h.p. Wright R-975 engine; for ser. #6B-2036 at 4230 lbs. gross wt. Refer to ATC #130 for more details.

#2-297 (11-5-30)

Bellanca CH-300 Spl. as 6PCLM with 300 h.p. Wright R-975 engine; for ser. #186 at 4300 lbs. gross wt.

#2-298 (11-5-30)

Keystone C4C as 6PCAmB with 525 h.p. Wright "Cyclone" engine; for ser. #600 at 6250 lbs. gross wt.

#2-299 (11-10-30)

Granville "Gee Bee" Sport C as 1POLM with 95 h.p. Menasco B-4 engine; for ser. #3 at 1335 lbs. gross wt. Refer to ATC #404 for similar examples.

#2-300 (11-14-30)

Travel Air 4D as 3POLB with 240 h.p. Wright R-760 engine; for ser. #1367 at 2880 lbs. gross wt. Refer to ATC #254 for similar examples.

#2-301 (11-15-30)

Stinson "Junior" SM-8A as 4PCLM with 215 h.p. Lycoming engine; for ser. #4201 at 3195 lbs. gross wt. Refer to ATC #295 for similar examples.

#2-302 (11-18-30)

Travel Air (Curtiss-Wright) S-6000-B Spl. as 4PCLM with 300 h.p. Wright R-975 engine; for ser. #779, 966, 1029 at 4230 lbs. gross wt. Refer to ATC #130 for the standard 6 place version.

#2-303 (11-18-30)

American Eagle "Eaglet" 230 as 2POLM with 30 h.p. Szekely SR-3L engine; for ser. #1000-1033 at 867 lbs. gross wt. See ATC # 380 (Vol. 4-249) for photo.

#2-304 (11-18-30)

Mercury "Mars" as 8PCLM with 420 h.p. Wasp B engine; for ser. #101 at 5700 lbs. gross wt. Very little is known about this airplane.

#2-305 (11-21-30)

Fleet 1 Spl. as 2POLB with 95 h.p. Menasco B-4 engine; for ser. #218 (NC-643M) at 1613 lbs. gross wt. Ship was modified into this configuration by Menasco Motors.

#2-306 (11-25-30)

Verville "Air Coach" 104 as 4PCLM with 240 h.p. Wright R-760 engine; for ser. #7 at 3400 lbs. gross wt. Refer to ATC #267 for similar examples.

#2-307 (11-29-30)

A. V. Roe "Avian" Spl. as 2POLB with 95 h.p. Menasco B-4 engine; for ser. #246 at 1525 lbs. gross wt. The Avro "Avian" was British-built.

#2-308 (11-1-30)

Travel Air (Curtiss-Wright) A-6000-A Spl. as 7PCLM with 420 h.p. Wasp C engine; for ser. #1095 at 5250 lbs. gross wt. See ATC #116 for similar examples.

#2-309 (12-11-30)

Buhl "Sport Airsedan" CA-3E as 4PCLB with 225 h.p. Packard Diesel DR-980 engine; for ser. #57 only. Refer to ATC #163 for more details.

#2-310 (12-16-30)

Boeing Tri-Motor 226 as 8PCLB with three 525 h.p. Hornet A-2 engines; for ser. #1091 only. See ATC #206 for more details.

#2-311 (12-18-30)

Waco RNF as 3POLB with 110 h.p. Warner engine; for ser. #3258 and 3393 only at 1900 lbs. gross wt. Refer to ATC #311 for similar examples.

#2-312 (12-26-30)

Swanson "Coupe" W-15 as 2PCLM with 110 h.p. Warner engine; for ser. #1 only at 2050 lbs. gross wt. Ser. #2-3 also eligible with certain restrictions.

#2-313 (12-26-30)

Stinson "Junior" SM-7AS as 4PCSM with 300 h.p. Wright R-975 engine; for ser. #3100 on Edo Q foats.

#2-314 (12-26-30)

Fairchild 71 Spl. as 5PCLM with 450 h.p. Wasp SC engine; for ser. #775 only.

#2-315 (12-26-30)

Zenith Z-6-B as 7PCLB with 420 h.p. Wasp C engine; for ser. #5 and up.

#2-316 (1-3-31)

Detroit-Lockheed "Vega" DL-1 Spl. as 5PCLM with 420 h.p. Wasp C-1 engine; for ser. #155 at 4500 lbs. gross wt. Del. to Comdr. Kidston of England. See ATC #308.

#2-317 (1-5-31)

Granville "Gee Bee" E as 1POLM with 110 h.p. Warner engine; for ser. #4 at 1400 lbs. gross wt. Refer to ATC #398 for similar examples.

#2-318 (1-6-31)

Viking "Kitty Hawk" B-8 as 3POLB with 125 h.p. Kinner B5 engine; for ser. #25 only. See ATC #392.

#2-319 (1-15-31)

Bellanca CH-300 Spl. as 6PCLM with 300 h.p. Wright R-975 engine; for ser. #180 only. See ATC #129.

#2-320 (1-20-31)

Savoia-Marchetti S-62 as 8PCFbB with 500 h.p. Issota-Fraschini "Asso" engine; for ser. #6202.

#2-321 (1-21-31)

Boeing Tri-Motor 80 Spl. as 10PCLB with three 450 h.p. Wasp engines; for ser. #1030 and 1032 only.

#2-322 (1-23-31)

Ford Tri-Motor 4-AT-D as 12-15PCLM with three 220 h.p. Wright J5 engines; for ser. #24 and 37 at 10,130 lbs. gross wt. Ser. #37 also had Wright R-975 engine in the nose & eligible as 16PCLM.

#2-323 (1-26-31)

Timm "Collegiate" TW-120 as 2POLM with 120 h.p. Western Enterprise L-7 engine; for ser. #106 and up.

#2-324 (2-7-31)

Laird LCB-200 (LC-2B-200) as 3POLB with 220 h.p. Wright J5 engine; for ser. #193, 195, and 198.

#2-325 (2-13-31)

Waterman W-1 as 4PCLM with 210 h.p. Kinner C5 engine; for ser. #1 only at 3269 lbs. gross wt. This a/c was called "Rubber Duck" because wings had adj. dihedral.

#2-326 (2-9-31)

New Standard D-29 Spl. as 2POLB with 95 h.p. Menasco B-4 engine; for ser. #1026 only.

#2-327 (2-12-31)

New Standard D-33 as 3POLB with 125 h.p. Kinner B5 engine; for ser. #33-1, 33-2, 33-3 at 2140 lbs. gross wt.

#2-328 (2-17-31)

Buhl "Sport Airsedan" CA -3CW as 3PCLB with 300 h.p. Wasp Jr. engine; for ser. #7 only.

#2-329 (3-2-31)

Hamilton Metalplane H-47 as 7PCLM with 525 h.p. Hornet A engine; for ser. #45, 48, 50, 52, 54, 56, 62, 63, and 69 at 6418 lbs. gross wt. See ATC #94 for similar examples.

#2-330 (3-4-31)

Stinson "Detroiter" SB-1 Spl. as 3PCLB with 200 h.p. Wright J4 engine; for ser. #1, 3, and 90.

#2-331 (3-9-31)

Consolidated "Fleetster" 17-2H Spl. as 5-6PCLM with 525 h.p. Hornet B-1 engine; for ser. #7 and 8 at 5600 lbs. gross wt. See ATC #291 for similar examples.

#2-332 (3-9-31)

Savoia-Marchetti S-56 as 2POAmB with 100 h.p. Kinner K5 engine; for ser. #5602 and 5604.

#2-333 (3-10-31)

Aeronca C-3 as 2POLM with 28 h.p. Aeronca E-107-A engine; for ser. #A-102 only. See ATC #396.

#2-334 (4-10-31)

American-Sunbeam "Pup" LP-1 as 2POLM with 40 h.p. Salmson (9 cyl.) AD-9 engine; for ser. #1 only.

#2-335 (4-14-31)

Stearman-Northrop "Alpha" 3 as 3PCLM with 420 h.p. Wasp engine; for ser. #4-8, 4-17, 4-18 and up.

#2-336 (4-11-31)

Buhl "Airsedan" CA-6J Spl. as 5PCLB with 300 h.p. Wasp Jr. engine; for ser. #56 only. Refer to ATC #128 for similar examples of the CA-6 series.

#2-337 (4-15-31)

Fleet "Model 9" as 2POLB with 125 h.p. Kinner B5 engine; for ser. #500 (NC-345N) only. Refer to ATC #428 (Vol. 5-84) for photo & additional data.

#2-338 (4-15-31)

Kreider-Reisner C6A as 2POLB with 110 h.p. Warner engine; for ser. #254X only at 1650 lbs. gross wt. Refer to ATC #215 for data and (Vol. 3-48) for additional photo. This airplane was the prototype for Fairchild KR-21 series.

#2-339 (4-15-31)

Great Lakes 2-T-1 Spl. as 2POLB with Menasco B-4 engine of 95 h.p.; for ser. #39, 69, 141 at 1580 lbs. gross wt. Refer to ATC #167 for 2-T-1 data.

#2-340 (4-21-31)

Ogden "Osprey" PC as 6PCLM with three 125 h.p. Menasco C-4 engines; for ser. #105 and up.

#2-341 (4-22-31)

Consolidated "Fleetster" 17-GW-1 as 8PCLM with 425 h.p. P & W R-1340-G engine; for ser. #1 only.

#2-342 (4-25-31)

Arrow Sport 66 "Tangerine" as 2POLB with 85 h.p. LeBlond 5DF engine; for ser. #801 thru 821.

#2-343 (5-4-31)

Chamberlin C-82 as 7PCLM with 300 h.p. Wright R-975 engine; for ser. #3 only at 4590 lbs. gross wt. Designed by trans-Atlantic flier Clarence Chamberlin.

#2-344 (5-31)

California "Cub" D-2 as 2POLM with 100 h.p. Cirrus engine; for ser. #1 only at 1592 lbs. gross.

#2-345 (5-13-31)

Robin Spl. as 3PCLM with 115 h.p. Tank V-470 or V-502 engine; for all ser. nos. under ATC 40-68.

#2-346 (5-19-31)

Laird "Speedwing" LCR-W450 as 2POLB with 420-450 h.p. Wasp C1 or SC1 engine; for ser. #162, 183 only at 3200 lbs. gross wt.

#2-347 (5-31)

Boeing "Monomail" 221A as 6-8PCLM with 575 h.p. Hornet B engine; for ser. #1153, 1154. See ATC #366.

#2-348 (5-20-31)

Great Lakes 2T-1A Spl. as 2POLB with 90 h.p. A.C.E. Cirrus engine; for ser. #202. See ATC #228 for data.

#2-349 (5-22-31)

Stewart M-2 as 5PCLM with two 300 h.p. Wright R-975 engines; for ser. #2 only at 5700 lbs. gross.

#2-350 (7-1-31)

Keystone "Patrician" K-78 as 14-18PCLM with three 525 h.p. Wright "Cyclone" engines; for ser. #205. Del. to Wright Aeronautical Corp. as NC-98N.

#2-351 (5-28-31)

Kellett "Autogiro" K-2 as 2POLAg with 165 h.p. Continental A-70 engine; for ser. #1 only.

#2-352 (6-2-31)

Cycloplane C-1 as 1POLM with 22 h.p. Cyclomotor A2-25 engine; for ser. #1-3 at 660 lbs. gross wt.

#2-353 (6-31)

Nicholas-Beazley NB-8G as 2POLM with 80 h.p. A-S "Genet" engine; for ser. #K-4 & a/c mfgd. prior 5-2-38 at 1175 lbs. gross wt. See ATC #452 for data.

#2-354 (6-9-31)

Keystone "Patrician" K-78-D as 20PCLM with three 525 h.p. Wright "Cyclone" engines; for ser. #206.

#2-355 (6-9-31)

Chamberlin C-81 as 8PCLM with 300 h.p. Wright R-975 engine; for ser. #2 only at 4950 lbs. gross wt.

#2-356 (6-11-31)

Sioux "Coupe" 90C as 2PCLM with 110 h.p. Warner engine; for ser. #402 only at 1580 lbs. gross wt.

#2-357 (6-12-31)

C-W Travel Air "Sportsman" 14-C as 3POLB with 185 h.p. Curtiss "Challenger" engine; for ser. #2001.

#2-358 (6-11-31)

Taylor "Cub" E-2 as 2POLM with 38 h.p. Continental A-40 engine; for ser. #12-25 at 925 lbs. gross wt. See ATC #455 (Vol. 5-162) for photo.

#2-359 (6-13-31)

Buckley LC-4 as 4PCLM with 300 h.p. Wasp Jr. engine; for ser. #1 only.

#2-360 (6-20-31)

Holloway HB-2 as 4POFbM with 180 h.p. Hispano-Suiza E engine; for ser. #H-2 only.

#2-361 (6-27-31)

Waco "Taper-Wing" JYM as 1-3POLB with 300-330 h.p. Wright R-975 engine; for ser. #J-3183, 3001, D-2, D-3 as mailplane for Northwest Airways at 3100 lbs. gross wt.

#2-362 (7-6-31)

Bird Cabin "Model E" as 1PCLB with 125 h.p. Kinner B5 engine; for ser. #6001 only. See ATC #419

#2-363 (7-8-31)

Waco GXE Spl. as 3POLB with 115 h.p. Tank V-470 or V-502 engine; for all a/c under ATC #13 if mod. by Milwaukee Parts Corp. at 2210 lbs. gross wt.

#2-364 (7-8-31)

Ireland "Neptune" N-2 as 4POAmB with 220 h.p. Wright J5 engine; for ser. #23 only. See ATC #153.

#2-365 (7-16-31)

Pilgrim 100-A as 10PCLM with 575 h.p. Hornet B engine; for ser. #6601. See ATC #443 for photo.

#2-366 (7-17-31)

Douglas "Dolphin" Model 1 Spl. as 10-12 PCAmM with two 300-330 h.p. Wright R-975 engines; for ser. #999 only at 8350 lbs. gross wt. Used by Wilmington-Catalina Airline.

#2-367 (7-17-31)

Lockheed "Orion" 9 as 3-5PCLM with 450 h.p. Wasp SC engine; for ser. #174. See ATC #421.

#2-368 (7-20-31)

Travel Air 2000-T as 3POLB with 115 h.p. Tank V-470 or V-502 engine; for all ser. nos. under ATC 30.

#2-369 (9-5-31)

Driggs "Skylark" 3-95A as 2POLB with 85 h.p. DH Gipsy engine; for ser. #19-20 at 1449 lbs. gross wt.

#2-370 (7-28-31)

Spartan C2-60 as 2POLM with 55 h.p. Jacobs L-3 engine; for ser. #J-1 only. See ATC #427.

#2-371 (7-29-31)

Northrop "Alpha" 4 as 7PO/CLM with 420 h.p. Wasp engine; for ser. #2. See ATC #451.

#2-372 (8-1-31)

Pilgrim KR-34 Spl. as 3POLB with 120 h.p. Moore (General Airmotors) engine; for ser. #329 only.

#2-373 (8-31)

Hayden-Clark-O'Day "Time Builder" W-6 as 1POLM with 40 h.p. Salmson AD-9 engine; for ser. #1 only at 925 lbs. gross wt.

#2-374 (8-10-31)

Lockheed "Sirius" 8-C "Sport Cabin" as 4PCLM with 420 h.p. Wasp C engine; for ser. #150. Del. to Bernarr McFadden the famous health-fad promoter.

#2-375 (8-14-31)

Aeronca "Duplex" C-3 (PC-3) as 2POL/SM with 36 h.p. Aernoca E-113 or E-113-A engine; for ser. #A-107 and A-154 on AP-A1900 twin-float gear. Refer to ATC #396 for data on C-3 series.

#2-376 (8-21-31)

Bach "Air Yacht" 3-CT-9K as 10PCLM with 420 h.p. Wasp B engine in the nose & two 210 h.p. Kinner C5 engines in the wings; for ser. #18 & 22 thru 30. Refer to ATC #271 and 299 for more data on 3-CT-9 series.

#2-377 (8-21-31)

Lockheed "Vega" 2D as 5PCL/SM with 300 h.p. Wasp Jr. engine; for ser. #38, 40, 58 at 3600 lbs. gross (land) & 4200 lbs. gross (sea) on AP-A9500 floats; ser. #40 was the seaplane. NC-197E was #38 and first had Wright J5 engine.

#2-378 (8-21-31)

Waco "Taper-Wing" CTO Spl. as 1POLB with 240 h.p. Wright R-760 engine; for ser. #A-151 & 3117 at 2471-2600 lbs. gross wt. Both airplanes had "I" wing struts & spl. streamlined landing gear; #3117 as 1-3POLB. See ATC #257.

#2-379 (8-28-31)

Great Lakes 2T-1E as 2POLB with 95 h.p. A.C.E. Cirrus Hi-Drive engine; for ser. #247. See ATC #354.

#2-380 (8-28-31)

Boeing "Flying Boat" 204A as 5PCFbB with 420 h.p. Wasp engine; for ser. #1077 only. See ATC #157.

#2-381 (31)

Travel Air B9-4000 as 1-3POLB with 300 h.p. Wright R-975 (J6-9-300) engine; for ser. #1001, 1010, 1103, 1307, 1396 and 1397. Had type B or C wings. These a/c were very popular for air-show work, and sky-writing.

#2-382 (9-3-31)

New Standard D-32 as 2POLB with 165 h.p. Wright R-540 engine; for ser. #3000 only at 2111 lbs. gross.

#2-383 (9-11-31)

Cain "Sport" CC-14 as 2POLM with 95 h.p. A.C.E. Cirrus Hi-Drive engine; for ser. #2 and up.

#2-384 (9-15-31)

Gaviota ST-1A as 1POLM with 65 h.p. Velie M-5 engine; for ser. #1 only.

#2-385 (9-22-31)

Lockheed "Vega" 5-C as 5PCSM with 420 h.p. Wasp engine; for ser. #55 on Edo W floats. See ATC #384.

#2-386 (10-17-31)

Lockheed "Altair" DL-2A as 3PCLM with 450 h.p. Wasp SC engine; for ser. #180. Detroit-Lockheed.

#2-387 (9-30-31)

American Eagle "Eaglet" 231 as 2POLM with 40 h.p. Salmson AD-9 engine; See ATC #450 for data & photo.

#2-388 (10-2-31)

Waco "Cabin" PDC as 4PCLB with 170 h.p. Jacobs LA-1 engine; for ser. #3512 only at 2558 lbs. gross wt.

#2-389 (10-7-31)

Taubman LC-13 as 2POLM with 75 h.p. Mich.-Rover engine; for ser. #601-602 only at 1450 lbs. gross wt.

#2-390 (10-24-31)

Emsco "Twin" B-5 as 8PCLM with two 300 h.p. Wright R-975 engines; for ser. #1 only. See ATC 400.

#2-391 (11-12-31)

Sikorsky S-39-C as 4-5PCAmM with 400 h.p. Wasp Jr. S1A engine; for ser. #911, 918 at 4014 lbs. gross wt. Ser. #920 also eligible, but as S-39-C Spl. with 400 h.p. Wasp Jr. TB or T3A engine. Refer to ATC #375 for basic data.

#2-392 (12-12-31)

Bellanca "Pacemaker" CH-400-W as 5PCLM with 400 h.p. Wasp Jr. S1A engine; for ser. #305, 307 only at 4277 lbs. gross wt.; was first a model 300-W on ATC #328.

#2-393 (12-16-31)

DeHaviland "Puss Moth" DH-80A as 3PCLM with 105 h.p. DH "Gipsy" 3 engine; for ser. #2140 at 1986 lbs. gross wt. Was British-built & imported to U.S.A.

#2-394 (12-21-31)

Davis D-1-W as 2POLM with 110-125 h.p. Warner engine; for ser. #504, 701-706, and 801. Refer to ATC #272 (Vol. 3-206) for photo of Walter Davis in cockpit of D-1-W.

#2-395 (1-4-32)

Fokker F-14-A as 10PCLM with 575 h.p. Hornet B engine; for ser. #1421 as modified from F-14 "Parasol." Sold to Canada as CF-AUD. See ATC #234 for std. F-14 data.

#2-396 (12-31-31)

Emsco "Sport" B-7-CH as 2POLM with 185 h.p. Curtiss "Challenger" engine; for ser. #1 only at 2100 lbs. gross wt. See ATC #424 for B-7-C series.

#2-397 (1-16-32)

Lockheed "Orion" 9A Spl. as 5PCLM with 450 h.p. Wasp SC engine; for ser. #187 to Hal Roach Studios.

#2-398 (1-28-32)

Boeing 40-B Mod. as 3PCLB with 525 h.p. Hornet A3 engine; for ser. #680. See ATC #27.

#2-399 (2-3-32)

Travel Air "Speedwing" B-11-D as 1POLB with 240 h.p. Wright R-760 engine; for ser. #1267 only at 2083 lbs. gross wt. This airplane used for air-racing.

#2-400 (2-4-32)

Lockheed "Sirius" 8 Spl. as 2POLM with 450 h.p. Wasp SC engine; for ser. #152 only for Victor Fleming.

#2-401 (2-10-32)

Stearman-Northrop "Beta" 3D as 1POLM with 300 h.p. Wasp Jr. engine; for ser. #2 only.

#2-402 (2-27-32)

Consolidated "Fleetster" 17-2CA as 6PCLM with 575 h.p. Wright R-1820-E engine; for ser. #10. See ATC #369 for data on 17-2C series.

#2-403 (3-9-32)

Curtiss-Wright Travel Air "Speedwing" B-14-R as 1-2POLB with 420 h.p. Wright R-975-E2 engine; for ser. #2003 at 2966 lbs. gross wt. Developed for "Casey" Lambert.

#2-404 (4-1-32)

Aeromarine-Klemm AKL-26 Spl. as 2POLM with 65 h.p. Velie M-5 engine; for (NC-102M) ser. #2-33. Refer to ATC #203 and 204 for data.

#2-405 (4-12-32)

Consolidated "Basic Trainer" 21-C as 2POLB with 300 h.p. Wasp Jr. A engine; for ser. #1 and 12 at 3100 lbs. gross wt. Several of these airplanes were exported for advanced stage military training.

#2-406 (4-14-32)

Stearman "Jr. Speedmail" 4DX as 2PCLB with 400 h.p. Wasp Jr. S1A engine; for ser. #4024 only at 3800 lbs. gross wt. Was fitted with sliding canopy enclosure.

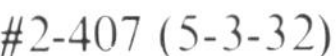

#2-407 (5-3-32)

Cessna AC as 4PCLM with 130 h.p. Comet engine; for ser. #150 only at 2260 lbs. gross wt.

#2-408 (5-3-32)

Spartan "Trainer" C2-165 as 2POLM with 165 h.p. Wright R-540 engine; for ser. #D-1 and D-2 only at 2140 lbs. gross wt. Used at Spartan School of Aeronautics for blind-flight and navigational training.

#2-409 (5-3-32)

Champion B-1 as 1POLM with 23 h.p. Cyclomotor A-2 engine; for ser. #101-102. An ultra-light solo-trainer.

#2-410 (5-6-32)

Barling B-6 as 6PCLM with 165 h.p. Continental A-70 engine; for ser. #A-1 at 3653 lbs. gross wt. Designed by Walter Barling.

#2-411 (5-25-32)

Boeing 95-A Spl. as 1POLB with 500 h.p. Wasp SD engine; for ser. #1068 & all Model 95 under ATC #106 when changed to conform.

#2-412 (6-7-32)

Boeing 203-B as 2POLB with 215 h.p. Lycoming R-680-B4 engine; for ser. #1138 and up. All 203-A a/c eligible when changed to conform. Modified by Boeing School of Aeronautics at Oakland, Calif. See ATC #211 (Vol. 3-39) for photo.

#2-413 (6-8-32)

Stinson Tri-Motor U as 10-11 PCLM with three 240 h.p. Lycoming R-680-BA engines; for ser. #9000 at 9300 lbs. gross wt. Refer to ATC #484.

#2-414 (6-11-32)

Stinson "Junior" SM-2K Spl. as 3PCLM with 125 h.p. Kinner B5 engine; for ser. #1012 only at 2520 lbs. gross wt. Refer to ATC #48 for SM-2 data.

#2-415 (6-30-32)

Amphibions "Privateer" P-3 as 2PCAmM with 165 h.p. Continental A-70 engine; for ser. #80 only.

#2-416 (7-11-32)

Lockheed "Orion" 9-C Spl. as 5PCLM with 450 h.p. Wasp SC1 engine; for ser. #180. Later converted to "Shelllightning" with 650-735 h.p. Wright SR-1820-F2 engine at 5824 lbs. gross wt. Flown regularly by "Jimmie" Doolittle.

#2-417 (7-14-32)

Caproni CA-100 as 2POLB with 85 h.p. DH "Gipsy" engine; for ser. #GC-C1. Imported from Italy.

#2-418 (7-15-32)

Waco UEC as 4PCL/SB with 210 h.p. Continental R-670 engine; for ser. #3636 and up on Edo P-3300 floats at 3250 lbs. gross wt. See ATC #467 for landplane version.

#2-419 (8-23-32)

Buhl "Airster" CA-1 as 1POLM with 300 h.p. Wasp Jr. engine; for ser. #59 as mailplane.

#2-420 (9-2-32)

Travel Air 6000 as 5PCSM with 220 h.p. Wright J5 engine; for ser. #1036 on Edo K floats. See ATC #100.

#2-421 (9-20-32)

Swallow F-27-W as 3POLB with 125 h.p. Warner "Scarab" engine; for ser. #847 only. A 1927 model.

#2-422 (10-10-32)

Bellanca CH-300 Spl. as 4PCLM with 300 h.p. Wright R-975 engine; for ser. #203 and 204.

#2-423 (10-8-32)

Lockheed "Altair" 8-D as 2PO/CLM with 450 h.p. Wasp SC1 engine; for ser. #143, 152, 214. Ser. #152 as 8-E had 550 h.p. Wasp S1D1 engine for Chas. Kingsford-Smith.

#2-424 (10-28-32)

Boeing Trainer 81-C as 2POLB with 165 h.p. Wright R-540 engine; for ser. #1037 only. This same airplane was previously an 81-A and 81-B.

#2-425 (10-29-32)

American Eagle 101-T as 3POLB with 115 h.p. Tank V-470 or V-502 engine; for ser. #329 at 2041 lbs. gross wt. The "Tank" was basically an OX-5 engine with air-cooled cylinder barrels.

#2-426 (11-2-32)

Kinner Sedan P as 2PCLM with 210 h.p. Kinner C5 engine; for ser. #1 only. Built for Robert Porter. Typical of the "Playboy" R-5.

#2-427 (11-9-32)

Lockheed "Vega" 1 Spl. as 5-7PCLM with 225 h.p. Packard Diesel DR-980 engine; for ser. #14 only. Converted by Clarence Chamberlin.

#2-428 (11-23-32)

Parks P-1-H as 3POLB with 100 h.p. Kinner K5 engine; for ser. #201-210 at 2010 lbs. gross wt. Became "Hammond 100" as per 1-23-33 on this approval.

#2-429 (11-28-32)

Fairchild 22-C7C as 2POLM with 120 h.p. DH "Gipsy" 3 engine; for ser. #901 to Canada. Refer to ATC #483 (Vol. 5-236) for photo.

#2-430 (12-1-32)

Waco ENF Spl. as 3POLB with 120 h.p. Martin 333 (Chevrolair) engine; for ser. #3379 only at 1905 lbs. gross wt. Converted by Glenn L. Martin Co.

#2-431 (12-2-32)

Kellett "Autogiro" K-2-A as 2POLAg with 210 h.p. Continental R-670 engine; for ser. #1-12 and 18 at 2265 lbs. gross wt. All K-2 models were eligible when changed to conform. Refer to ATC #437 for K-2 data.

#2-432 (12-28-32)

Travel Air 4-U as 3POLB with 130-165 h.p. Comet engine; for ser. #242, 268, 322, 418, 423, 551, 611, 735, 763, 767, 788, 818, 871, 1004, 1016, 1162. Remodeled by Otto W. Timm who held this approval.

#2-433 (12-20-32)

Douglas O2-MC4 as 2POLB with 450 h.p. Wasp SC1 engine; for ser. #1162 thru 1173.

#2-434 (12-29-32)

Sikorsky S-38-BS as 7PCAmB with two 450 h.p. Wasp SC1 engines; for ser. #414-20 only at 10,480 lbs. gross wt. See ATC #126 for S-38-B data.

#2-435 (12-29-32)

Fokker F-14-B as 8-10PCLM with 575 h.p. Hornet B-1 engine; for ser. #1412. Exported to Costa Rica.

#2-436 (12-29-32)

Sikorsky S-39-CS as 4PCAmM with 375 h.p. Wasp Jr. S2A engine; for ser. #914 at 4013 lbs. gross wt.

#2-437 (2-7-33)

Stinson Tri-Motor U-1 as 11PCLB with three 285 h.p. Wright R-760-E1 engines; for ser. #9014 only.

#2-438 (2-15-33)

States B-3 as 2POLM with 100 h.p. Kinner K5 engine; for ser. #111 as assembled by Adam Bialorski. Refer to ATC #349 for B-3 data.

#2-439 (2-21-33)

DeHaviland "Fox Moth" DH-83 as 3-4PCLB with 122 h.p. DH Gipsy Major engine; for ser. #4026 at 2070 lbs. gross wt. Imported to the U.S.A.

#2-440 (2-25-33)

Command-Aire 3C3-B as 3POLB with 113 h.p. Siemens-Halske SH-14 engine; for ser. #W-145. See ATC #120.

#2-441 (3-3-33)

Bird model CC as 3POLB with 185 h.p. Curtiss "Challenger" engine; for ser. #9001 only at 2336 lbs. gross wt. Operated by Curtiss-Wright Flying Service. Refer to ATC #419 for additional data.

#2-442 (3-11-33)

Curtiss-Wright "Junior" CW-1A as 2POLM with 35 h.p. Augustine B4-40 engine; for ser. #1225 only. Refer to ATC #397 for standard "Junior" data.

#2-443 (4-20-33)

Fleet Model 9 as 2POSB with 125 h.p. Kinner B5 engine; for ser. #510 only on Edo I twin-float gear at 1956 lbs. gross wt. Refer to ATC #428 (Vol. 5-86) for photo.

#2-444 (4-29-33)

Westbrook "Sportster" W-5 as 2POLM with 100 h.p. A.C.E. Cirrus Mk. 3 engine; for ser. #W-504 only at 1500 lbs. gross wt. Later appearing on some approval listings as the Continental C-2 & then as Allied H-28.

#2-445 (5-1-33)

Stearman "Speedster" C3P as 2POLB with 220 h.p. Wright "Whirlwind" J5 engine; for ser. #5039 at 2830 lbs. gross wt. This airplane is a C3R type as featured on ATC #251; it was apparently the last C3R airplane built.

#2-446 (5-18-33)

Sikorsky "Amphibion" S-38-BT as 10PCAmB with two 525 h.p. supercharged Wasp T1D1 engines; for ser. #314-20 and 414-8. Ser. #414-13 also eligible, but as 8PCAmB. Refer to ATC #126 for detailed S-38-B data. These three airplanes were built in the 3rd and 4th block of production quotas.

#2-447 (5-19-33)

Breese R-6-C as 5PO/CLM with 330 h.p. Wright R-975-E engine; for ser. #10 only at 4000 lbs. gross wt. Built as business-plane for Russell H. Lawson; designed by Vance Breese, famous contract test-pilot.

#2-448 (5-29-33)

Detroit-Lockheed "Vega" DL-1B Spl. as 7PCLM with 450 h.p. Wasp SC1 engine; for ser. #161 only at 4750 lbs. gross wt. Assembled by Richard A. Von Hake from left-over parts at the Burbank & Detroit plants.

#2-449 (6-1-33)

Amphibions "Privateer" P3B as 3PCAmM with 210 h.p. Continental R-670 engine; for ser. #81 thru 84 at 3200 lbs. gross wt. Amphibions, Inc. was formerly Ireland Aircraft, Inc.

#2-450 (6-16-33)

Walden-Markey WM-1 as 4POFbB with 180 h.p. Hisso (Hispano-Suiza) E engine; for ser. #10 only.

#2-451 (6-29-33)

Security-National "Airster" S-1 as 2POLM with 100 h.p. Kinner K5 engine; for ser. #10 and 11 only at 1775 lbs. gross wt. A "Bert" Kinner design.

#2-452 (7-1-33)

Monocoupe 110 Spl. as 2PCLM with 125-145 h.p. Warner engine; for ser. #5W40, 5W47 and up at 1875 lbs. gross wt. The original a/c was remodeled by John H. Wright for air-racing. See ATC #327 (Vol. 4-98) for photo.

#2-453 (7-6-33)

Bach Transport T-11-P as 10PCLM with 525 h.p. Hornet A engine; for ser. #10, 16, 24, 25, 26, 27.

#2-454 (7-8-33)

Verville "Air Coach" 104-A as 4PCLM with 220 h.p. Wright J5 engine; for ser. #8 only. Was formerly a model 104-C as NS-11. Refer to ATC #267.

#2-455 (7-20-33)

Westbrook "Sport" W-5-B as 2POLM with 85 h.p. British Cirrus Mk. 3 engine; for ser. #W-503.

#2-456 (8-5-33)

Northrop "Delta" 1-A as 1-7PCLM with Wright SR-1820-F3 engine of 710 h.p.; for ser. #3 as NC-12292. This was the prototype "Delta."

#2-457 (8-18-33)

Stearman "Cloudboy" 6-C as 2PO/CLB with 330 h.p. Wright R-975-C engine; for ser. #6010 only at 3047 lbs. gross wt. Modified from "Stearman 6."

#2-458 (8-18-33)

Northrop "Delta" 1-B as 9PCLM with 660 h.p. Hornet T2C1 engine; for ser. #4 to Aerovias Centrales.

#2-459 (9-8-33)

Bellanca "Skyrocket" F-1 as 6PCLM with 450 h.p. Wasp SC1 engine; for ser. #802 only.

#2-460 (11-8-33)

Douglas DC-1A as 14PCLM with two 700 h.p. Hornet SDG engines; for ser. #1137 only. Also eligible with two 710 h.p. wright SGR-1820-F3 engines.

#2-461 (11-8-33)

Stinson "Junior" SM-8A Spl. as 4PCLM with 220 h.p. Wright J5 engine; for ser. #4014 only at 3195 lbs. gross wt. Robt. W. Ellis as remodeler. See ATC #295.

#2-462 (11-11-33)

Bellanca "Skyrocket" F-2 as 6PCLM with 450 h.p. Wasp SC1 engine; for ser. #802-803 at 5600 lbs. gross wt.

#2-463 (11-23-33)

Viking "Kitty Hawk" B-8 as 3POSB with 125 h.p. Kinner B5 engine; for ser. #26 and 31 on L-2260 Edo floats at 2270 lbs. gross wt. See ATC #392 (Vol. 4-287) photo.

#2-464 (12-1-33)

Lockheed "Orion" 9D-1 as 5-7PCLM with 550 h.p. Wasp S1D1 engine; for ser. #204 to John Mabee. This airplane had wing flaps.

#2-465 (1-4-34)

National "Bluebird" C-3 (LP-1) as 2POLM with 40 h.p. Salmson AD-9 engine; for ser. #3 at 925 lbs. gross wt. Later had "National" engine as shown. Patterned after American-Sunbeam "Pup" (#2-334) by National Airplane & Motor Co. of Billings, Montana.

#2-466 (2-2-34)

Uppercu (Aeromarine-Klemm) AKL-26A as 2POSM with 65 h.p. LeBlond 5D engine; for ser. #2-51 only on Kantner floats. Refer to ATC #204 for AKL-26A data.

#2-467 (2-14-34)

Uppercu (Aeromarine-Klemm) AKL-26B as 2POLM with 85 h.p. LeBlond 5DF engine; for ser. #87 only. See ATC #334 for AKL-26B data.

#2-468 (2-15-34)

Northrop "Delta" 1-C as 8PCLM with 700 h.p. Hornet T1C1 engine; for ser. #7. Exported as SE-ADI.

#2-469 (2-20-34)

Driggs "Skylark" model 3 as 2POLB with 75 h.p. Michigan-Rover engine; for ser. #3020 only. Same as airplanes on ATC #303 which had by now expired & company gone out of business. Assembled by Skylark Aircraft Co.

#2-470 (2-24-34)

Comper "Swift" series C-7 as 1POLM with 105 h.p. DeHaviland "Gipsy" Mk. 3 engine; for ser. #GS-32-2 at 1130 lbs. gross wt. Imported from England. The "Swift" was a very popular sportplane in the British Isles.

#2-471 (2-28-34)

Stinson "Reliant" SR Spl. as 4PCLM with 240 h.p. Lycoming R-680-BA engine; for ser. #8788 only at 3325 lbs. gross wt. Refer to ATC #510 for detailed SR data. This was the first of the famous "Reliant" series.

#2-472 (3-1-34)

Curtiss "Fledgling" J-1 Spl. as 2POLB with 220 h.p. Wright J5 engine; for ser. #B-22 at 2700 lbs. gross wt. and ser. #B-69 at 2962 lbs. gross wt.

#2-473 (3-13-34)

Clyde V. Cessna C-3 as 4PCLM with 125 h.p. Warner "Scarab" engine; for ser. #4 only at 2260 lbs. gross wt. Built for Walt Anderson originator of "White Castle" hamburger stands which were popular in the midwest.

#2-474 (34)

Welch OW-5M as 1-2POLM with 37 h.p. Continental A-40 engine; for ser. #108 only at 954 lbs. gross wt. Refer to ATC #636 (Vol. 7-132) for photo. (-13500).

#2-475 (6-5-34)

Bellanca "Skyrocket" F as 5-6PCLM with 550 h.p. Wasp S1D1 engine; for ser. #804-805-806 at 5600 lbs. gross wt. Ser. #805 as shown was with Tennessee Valley Authority (TVA). Ser. #805-806 had wing flaps; #804 did not.

#2-476 (6-8-34)

Northrop "Delta" 1-E as 2PCLM with 660 h.p. Hornet T2D1 engine; for ser. #29 only. First as NR-13755 then shipped to Sweden.

#2-477 (6-34)

General Mfg. GA-43J as 12PL/SM with 700 h.p. Hornet T1D1 engine; for ser. #2205 on Edo 36-9225 twin-float gear. Del. to SCADTA (Colombia) on 11-19-34. See ATC #527 (Vol. 6-89) for photo & applicable data. Entered as North American GA-43J on some listings.

#2-478 (6-26-34)

Straughan A as 2POLB with 40 h.p. Straughan AL-1000 (Ford) engine; for ser. #2. Refer to ATC #561 (Vol. 6-219) for photo & applicable data. Later became "Wiley Post."

#2-479 (7-5-34)

Douglas "Dolphin" 11 as 5PCAmM with two 450 h.p. Wasp S3D1 engines; for ser. #1278 only. This was the deluxe "Executive" model.

#2-480 (7-11-34)

Kinner "Playboy" R as 2PCLM with 160 h.p. Kinner R5 engine; for ser. #84 at 2200 lbs. gross wt. See ATC #518 (Vol. 6-64) for photo & pertinent data.

#2-481 (7-16-34)

Keystone-Loening "Commuter" K-84 Spl. as 2PCAmB with 365 h.p. Wright R-975-E1 engine; for ser. #337 only. Refer to ATC #219 for data on std. K-84.

#2-482 (7-30-34)

Douglas "Dolphin" 8-117 as 7PCAmM with two 450 h.p. Wasp S3D1 engines; for ser. #1280 and up mfgd. prior to 7-1-40 at 9500 lbs. gross wt. Ser. #1280 del. to Wm. E. Boeing as his private airplane named the "Rover."

#2-483 (8-3-34)

Douglas "Dolphin" 8-114 as 13PCAmM with two 450 h.p. Wasp SC1 or S1C1 engines; for ser. #1279 & 1282 at 9530 lbs. gross wt.

#2-484 (8-3-34)

Northrop "Delta" 1-D as 5PCLM with 710 h.p. Wright SR-1820-F3 engine; for ser. #28 only at 7350 lbs. gross wt. See ATC #553 (Vol. 6-193) for photo. Operated by Richfield Oil Co.

#2-485 (8-16-34)

Ford "Freighter" 8-AT and 8-ATS as 2-13PCL/SM with 700 h.p. Wright "Cyclone" GR-1820-F1 engine; for ser. #8-AT-1 only. As shown, this airplane operated in Alaska as landplane or seaplane.

#2-486 (8-27-34)

Douglas "Dolphin" 12 as 5-7PCAmM with two 400 h.p. Wasp Jr. SB engines; for ser. #1283 and 1284 at 9500 lbs. gross wt.

#2-487 (8-30-34)

Douglas "Dolphin" 129 as 10PCAmM with two 450 h.p. Wasp S6D1 engines; for ser. #1348-1349 only at 9500 lbs. gross wt. Without landing gear these airplanes were classed as 8PCFbM.

#2-488 (9-11-34)

Lockheed "Orion" 9D-2 as 4-5PCLM with 550 h.p. Wasp S1D1 engine; for ser. #208 only at 5800 lbs. gross wt. Used regularly by "Detroit News" for gathering news and photos; flown by James V. Piersol. As UC-85 during World War II.

#2-489 (9-29-34)

Northrop "Gamma" 2-G as 2PCLM with 745 h.p. Curtiss SVG-1570-F4 (supercharged) engine; for ser. #11 only. Flown by "Jackie" Cochrane.

#2-490 (10-2-34)

Northrop "Delta" 1-D3 as 6PCLM with 650 h.p. Hornet S4D1 engine; for ser. #40 only as NC-14265.

#2-491 (10-4-34)

Stinson "Reliant" SR-5A as 4PCL/SM with 245 h.p. Lycoming R-680-6 engine; for ser. #9297-A on Edo floats with Maine Dept. of Forestry. Refer to ATC #536.

#2-492 (10-4-34)

Stinson "Reliant" SR-5 Spl. as 4PCLM with 225 h.p. Lycoming R-680-4 engine; for ser. #9200-A and 9201-A at 3275 lbs. gross wt. Same as SR-5 except had wing flaps & small tail group. Refer to ATC #530.

#2-493 (10-4-34)

Stinson "Reliant" SR-5D as 4PCLM with 215 h.p. Lycoming R-680 engine; for ser. #9211 only at 3241 lbs. gross wt. This a/c had small tail-group. Refer to ATC #531.

#2-494 (10-4-34)

Stinson "Reliant" SR-5 Spl. as 4PCLM with 245 h.p. Lycoming R-680-6 engine; for ser. #9202, 9205, and 9210 at 3475 lbs. gross wt.

#2-495 (12-7-34)

Douglas DC-2A as 15-16PCLM with two 700 h.p. Hornet SDG engines; for ser. #1328 only. Del. to Standard Oil Co. of Calif. See ATC #570 (Vol. 6-251) for photo.

#2-496 (12-31-34)

Stinson "Reliant" SR-1 Spl. as 4PCLM with 225 h.p. Lycoming R-680-B4 engine; for ser. #8900 only at 3281 lbs. gross wt. Del. to Lucius B. Manning.

#2-497 (1-9-35)

Curtiss-Courtney CA-1 as 5PCAmB with 365 h.p. Wright R-975-E1 engine; for ser. #101. See ATC #582 (Vol. 6-293) for photo and pertinent data.

#2-498 (2-21-35)

Porterfield 35 and 35-70 as 2PCLM with 65 h.p. LeBlond 5D or 70 h.p. 5DE engines; for ser. #101 thru 131 and 133-134, 139 were eligible at 1235 lbs. gross wt.

#2-499 (3-6-35)

Stinson "Reliant" SR Spl. as 4PCLM with 215 h.p. Lycoming R-680 engine; for ser. #8756. See ATC #510.

#2-500 (35)

Swallow TP-K as 2POLB with 100 h.p. Kinner K5 engine; for ser. #213. Refer to ATC #186.

#2-501 (4-1-35)

Laird LC-1B-285 as 2-3POLB with 285 h.p. Wright R-760-E1 engine; for ser. #199 only at 2909 lbs. gross wt. By this time Laird production had just about ended.

#2-502 (5-35)

Bird model RK as 2POLB with 160 h.p. Kinner R5 series 2 engine; for ser. #CK-4039 only at 2250 lbs. gross wt. Converted by Kinner Airplane & Motor Co.

#2-503 (5-23-35)

Northrop "Gamma" 2-EDC (2-E) as 2PO/CLM with 710 h.p. Wright R-1820-F3 engine; for ser. #47 (NC-13760).

#2-504 (5-28-35)

Ford Tri-Motor 5-AT-DS as 12-13PCSM with three 450 h.p. Wasp S1D1 engines; for ser. #114.

#2-505 (6-4-35)

Fairchild KR-34-B2 as 3POLB with 165 h.p. Continental A-70-2 engine; for ser. #180 only at 2401 lbs. gross wt. Still being flown by Wm. E. Clark in Penna.

#2-506 (6-29-35)

Parks P-1-T as 3POLB with 115 h.p. Tank V-502 engine; for ser. #18-19 at 2078 lbs. gross wt., and all a/c under ATC #179 when changed to conform. Remodeled by Milwaukee Parts Co.

#2-507 (7-9-35)

Douglas DC2-115F as 16PCLM with two Bristol "Pegasus" series 6 engines rated 690 h.p.; for ser. #1377 and 1378. The "Pegasus" were British engines.

#2-508 (7-11-35)

Bellanca "Aircruiser" 66-70 as 1-12PCL/SM with 660 h.p. Wright SGR-1820-F32 engine; for ser. #719 on twin-float gear. See ATC #563 (Vol. 6-227) for photo.

#2-509 (7-20-35)

Bellanca "Pacemaker" 31-42 as 2-8PCL/SM with 420 h.p. Wright R-975-E2 engine; for ser. #251-252 only at 5600 lbs. gross (passenger) & 5900 lbs. gross (cargo). Same as ATC #578 but had Canadian-built floats.

#2-510 (7-23-35)

Fairchild KR-31-A as 1POLB with 100 h.p. Kinner K5 engine; for ser. #148 only at 1341 lbs. gross wt.

#2-511 (7-25-35)

Stinson "Reliant" SR-5A Spl. as 4PCLM with 245 h.p. Lycoming R-680-6 engine; for ser. #9349-A at 3550 lbs. gross wt. Operated by New York News Syndicate.

#2-512 (8-5-35)

Waco WHD as 2PO/CLB with 440 h.p. Wright R-975-E3 engine; for ser. #3837 at 3800 lbs. gross wt.

#2-513 (8-8-35)

Kellett "Autogiro" KD-1 as 2POLAg with 225 h.p. Jacobs L-4 engine; for ser. #102 and up. Refer to ATC #712 for data on KD-1 series.

#2-514 (8-28-35)

Rose "Parrakeet" A-1 as 1POLB with 40 h.p. Continental A-40-4 engine; for ser. #101 and up at 728 lbs. gross wt.

#2-515 (9-3-35)

Corben "Junior Ace" 6 as 2PCLM with 40 h.p. Salmson AD-9 engine; for ser. #134 and up at 1035 lbs. gross wt., mfgd. prior to 7-1-40.

#2-516 (9-23-35)

Stearman 70 as 2POLB with 215 h.p. Lycoming R-680 engine; for ser. #7001 only at 2750 lbs. gross wt. Progenitor of the Model 75 (Kaydet) series.

#2-517 (10-30-35)

North American NA-16 & NA-18 as 2PCLM with 550 h.p. Wasp S1D1 engine; for ser. #NA-16-1 & NA-18-1 at 4500 lbs. gross wt. An NAA development for the USAF.

#2-518 (11-4-35)

Ford Tri-Motor 5-AT-CS as 16PCSM with three 420 h.p. Wasp engines; for ser. #69 on twin-float gear.

#2-519 (1-9-36)

Bellanca "Aircruiser" 66-67 as 12PCLM with 675 h.p. Wright SR-1820-F32 engine; for ser. #704 at 9590 lbs. gross wt. Later #720 was also included.

#2-520 (2-12-36)

Stearman "Cloudboy" 6-P as 2POLB with 220 h.p. Wright J5 engine; for ser. #6002 only at 2500 lbs. gross wt. Was formerly a Model 6-A.

#2-521 (5-19-36)

General "Aristocrat" 102 as 3PCLM with 125 h.p. Warner engine; for ser. #3 only at 2000 lbs. gross.

#2-522 (8-6-36)

Ryan STA Spl. as 1POLM with 150 h.p. Menasco C4S engine; all a/c eligible when changed to conform.

#2-523 (9-3-36)

Rearwin 9000 & 9000 Deluxe as 2PCLM with 90 h.p. Warner Jr. engine; same as ATC #624.

#2-524 (10-15-36)

Miles "Sparrowhawk" M-5 as 1POLM with 130 h.p. DH Gipsy Major HC engine; for ser. #264 only at 1700 lbs. gross wt. Imported from England by Phillips & Powis.

#2-525 (9-29-36)

Curtiss-Wright "Junior" CW-1S as 2POLM with 40 h.p. Salmson AD-9 engine; for ser. #1164 & 1224 at 988 lbs. gross wt. Remodeled by Franklin R. Hyde.

#2-526 (11-10-36)

Keystone-Loening "Commuter" K-84-W as 4PCAmB with 330 h.p. Wasp Jr. TB engine; for ser. #313 only at 4270 lbs. gross wt. Remodeled by E. H. Hunt.

#2-527 (11-30-36)

Bird AT as 3POLB with 115 h.p. Tank V-502 engine; for all a/c under ATC #101 when changed to conform at 2150 lbs. gross wt. Approval held by Milwaukee Parts Co.

#2-528 (10-14-36)

Luscombe "Phantom" 1 & 1S as 2PCL/SM with 145 h.p. Warner engine; superseded by ATC #552.

#2-529 (12-17-36)

Taylorcraft A as 2PCLM with 40 h.p. Continental A-40-4 engine; approval superseded by ATC #643.

#2-530 (2-16-37)

Porterfield "Zephyr" CP-40 and CP-40A as 2PCLM with 40 h.p. Continental A-40-4 engine; for ser. #250 and up at 1040 lbs. gross wt., mfgd. prior to 7-1-40.

#2-531 (3-9-37)

Arrow "Sport" F as 2PO/CLM with 82 h.p. Arrow V-8 (Ford) engine; for all a/c mfgd. between 3-9-37 and 7-1-40 at 1675 lbs. gross wt. Otherwise same as ATC #613.

#2-532 (3-16-37)

Davis V-3 Spl. as 2POLM with 90 h.p. LeBlond 7D engine; for ser. #101 only at 1373 lbs. gross wt. Remodeled by M.E. Stone.

#2-533 (3-24-37)

Taylor "Cub" J-2 as 2POLM with 40 h.p. Continental A-40 engine; superseded by ATC #595.

#2-534 (4-14-37)

Fairchild KR-31-B as 2PCLB with 100 h.p. Kinner K5 engine; for ser. #213 only at 1729 lbs. gross wt. Remodeled by August Graf. Also as 1POLB.

#2-535 (4-20-37)

Northrop "Gamma" 2D2 as 2PO/CLM with 710 h.p. Wright SR-1820-F3 engine; for ser. #12 only at 7350 lbs. gross wt. NR-2111 for Russell Thaw. Was first on ATC #553.

#2-536 (5-26-37)

Stearman X75L3 as 2POLB with 225 h.p. Lycoming R-680-B4B engine; for ser. #X75-00-1 at 2675 lbs. gross wt. This airplane was prototype for "Kaydet" series.

#2-537 (10-30-37)

Martin 156 as 53PCFbM with four 1000 h.p. Wright GR-1820-G2 engines; for ser. #714 at 63,000 lbs. gross wt.; also eligible on ATC #697. See ATC #697 (Vol. 7-335) for photo. This airplane was shipped to Russia.

#2-538 (6-15-37)

Swallow C as 2PCLM with 125 h.p. Menasco C-4 engine; for ser. #C-1105 and C-1106 at 1980 lbs. gross.

#2-539 (7-8-37)

Vultee V1A Spl. as 8PCLM with 1000 h.p. Wright SR-1820-G5 engine; for ser. #25 at 8600 lbs. gross wt. Ser. #18 eligible with 890 h.p. Wright R-1820-F32 engine. Same as ATC #545 except changes.

#2-540 (8-6-37)

Fleetwings "Sea Bird" F-401 as 4PCAmM with 285 h.p. Jacobs L-5 engine; for ser. #401 only at 3430 lbs. gross wt. This airplane is still flying.

#2-541 (11-9-37)

Monocoupe G as 2PCLM with 90 h.p. Lambert R-266 engine; for ser. #G-1 thru G-5; later became "Dart." See ATC #674 (Vol. 7-257) for photo.

#2-542 (10-30-37)

New Standard (Jones) D-25 as 5POLB with 220 h.p. Wright J5 engine; for all a/c mfgd. by Ben Jones between 6-1-35 and Feb. 1938 at 3400 lbs. gross wt. See ATC #108.

#2-543 (4-4-38)

Consolidated 28 (PBY-2) as 10-14 PCFbM with two 900 h.p. Twin-Wasp SBG engines; for ser. #C-3 only at 27,000 lbs. gross wt. This a/c was the famous "Guba."

#2-544 (6-10-38)

Grumman G-32 and G-32A as 2PCLB; for ser. #446 (G-32) with Wright R-1820-G2 engine of 1000 h.p. & ser. #447 (G-32A) with 890 h.p. SR-1820-F52 engine at 4663 lbs. & 4372 lbs. gross respectively. "Gulfhawk" was flown by Al Williams.

#2-545 (8-18-38)

Clark (Fairchild) Duramold F-46-A as 5PCLM with 450 h.p. Ranger SGV-770-B5 engine; for ser. #5000 and up at 4800 lbs. gross wt. "Duramold" was all-wood construction.

#2-546 (10-31-37)

Pasped "Skylark" W-1 as a 2PO/CLM with 125 h.p. Warner engine; for ser. #1 only at 1900 lbs. gross wt. This a/c reflects the design-style of this period.

#2-547 (10-31-37)

Akron-Funk B as 2PCLM with 65 h.p. Akron-Funk E engine; for ser. #B-2 only at 1350 lbs. gross wt. Refer to ATC #715 for similar airplanes.

#2-548 (6-29-39)

Consolidated 28-4 as 4-6PCFbM with two 1050 h.p. Twin-Wasp SC3G or S1C3G engines; for ser. #C-4 only at 27,000 to 30,500 lbs. gross wt. Similar to a/c on ATC #785. Army and Navy models also eligible when modified to conform.

#2-549 (7-17-39)

Curtiss-Wright A-22 and A-22B as 2PCLM with 450 h.p. Wright R-975-E3 engine; for ser. #A-22-1 & 3604 at 3455 lbs. gross wt. SNC-1 (Navy) also eligible when modified.

#2-550 (8-22-39)

Meyers OTW as 2POLB with 125 h.p. Warner engine; for ser. #1 only at 1770 lbs. gross wt. Refer to ATC #736 (Vol. 8-133) for photo.

#2-551 (5-5-39)

Douglas DC-4E as 50PCLM with four 1150-1400 h.p. Twin-Hornet S1A1G engines; for prototype a/c only at 60,000 lbs. gross wt. See ATC #762 (Vol. 8-216) for photo & specs. Because of no airline interest, a/c was sold to Japan.

#2-552 (10-1-37)

Bennett (Globe) BTC-1 as 7PCLM with two 300 h.p. Jacobs L-5 engines; for ser. #1 only at 6745 lbs. gross wt. This was an all-wood airplane of "plastic" construction.

#2-553 (10-9-39)

Douglas (Northrop) 2J2 as 3PO/CLM with 550 h.p. Wasp S3H1G engine; for ser. #186 only at 6648 lbs. gross wt. This was a "Gamma" mfgd. at El Segundo Div. of Douglas Aircraft Col. Considered the progenitor of Army's A-17.

#2-554 (10-19-39)

Douglas O-38-B as 2POLB with 525 h.p. Hornet SD-1 engine; for ser. #31-411 and 31-433 only at 5000 lbs. gross wt. Remodeled by Civil Aeronautics Authority.

#2-555 (4-23-40)

Boeing F4B4 as 1PO/CLB with 550 h.p. Wasp S1D1 engine; for ser. #9241 and 9251 only at 3550 lbs. gross wt. Remodeled by Air Safety Board for CAA use.

#2-556 (4-30-40)

Lockheed "Lodestar" 18-H as 17PCLM with two 900 h.p. Hornet S1E2G engines; the eligible ser. nos. are unknown, records were not that explicit.

#2-557 (40)

New Standard (White) D-25-B as 1POLB with 285 h.p. Wright R-760-E1 engine; for ser. #159W. 164W-167W only at 3400 lbs. gross wt. Same as a/c on ATC #110 except engine; mfgd. by White Aircraft Corp.

#2-558 (10-28-40)

Interstate (Harlow) "Cadet" S-1 as 2PCLM with 50 h.p. Continental A-50-8 engine; for ser. #1 only at 1150 lbs. gross wt. See ATC #737 for similar airplanes.

#2-559 (3-3-41)

Superior (Culver) LAR as 1PCLM with 80 h.p. Franklin 4AC-176-F3 engine; for ser. #L-101-AR only at 1305 lbs. gross wt. A Culver design sold to Superior Aircraft.

#2-560 (8-11-41)

Travel Air L-4000 as 2POLB with 225 h.p. Lycoming R-680-B4 engine; for ser. #1122, 1180 at 2450 lbs. gross wt. Remodeled by Parks Air College and used in their training system. Additional a/c of the 4000 type may be converted.

#2-561 (41)

Jan Herse RWD-13 series 3 as 3PCLM with 120 h.p. DH Gipsy Major Mk. 1 engine; for ser. #158-159 only at 1962 lbs. gross wt. Believed to be a "Puss Moth."

#2-562 (3-10-42)

Phillips-Fleet 7 as 2POLB with 120 h.p. Martin-Phillips 333 engine; for ser. #212 and 403 at 1678 lbs. gross wt. Remodeled by Phillips Aviation Co.

#2-563

Piper "Cub Cruiser" J5C as 3PCLM with 100 h.p. Lycoming O-235 engine; same as ATC #725.

#2-564

Funk C as 2PCLM with 75 h.p. Continental A-75-8 engine; for ser. #48 only at 1350 lbs. gross wt. Remodeled by Ballauer & Chapper. See ATC #715 for similar a/c.

#2-565

No information was available for this approval.

#2-566

Fleet 16-B (RCAF "Finch") as 2PO/CLB with 125 h.p. Kinner B5 engine; for all ser. nos. with Canadian Airworthiness Certificate at 1878 lbs. gross wt. Over 600 built in Canada by Fleet Aircraft, Ltd. to RCAF for training WW2 pilots.

#2-567

Driggs "Skylark" 3-95 as 2POLB with 95 h.p. A.C.E. Cirrus Hi-Drive engine; for ser. #3016 only. Remodeled by Skylark Aircraft Corp. Formerly had "Rover" engine.

#2-568

Morrow "Sport" 1-L as 2PCLM with 185 h.p. Lycoming O-435-C engine; for ser. #2 only at 2350 lbs. gross wt. Rights turned over to Harlow Aircraft.

#2-569 (8-11-44)

Naval Aircraft Factory N3N-3 as 2POLB with 235 h.p. Wright R-760-8 engine; for all ser. nos. Some 816 a/c del. to U.S. Navy.

#2-570 (45)

Taylorcraft TG-6 (glider) as 2PCLM with 65 h.p. Continental A-65 engine; converted by Commonwealth Aircraft Corp.

#2-571 (5-31-45)

Consolidated-Vultee BT-13 (BT-15) as 2PCLM with 450 h.p. R-985 engine (BT-13) & 420 h.p. Wright R-975-E3 engine (BT-15); for all ser. nos. SNV also eligible.

#2-572 (8-1-45)

North American O-47-B as 1-3PCLM with 1100 h.p. Wright GR-1820-G102 engine; for all ser. nos. at 8380 lbs. gross wt.; 74 a/c del. to USAAF.

#2-573 (11-19-45)

Timm N2T1 as 2POLM with 220 h.p. Continental W-670-6A engine; for all U.S. Navy ser. nos. See ATC #750 (Vol. 8-179) for photo.

#2-574 (10-15-45)

Canadian-Vickers "Stranraer" as 3PCFbB with two Bristol "Pegasus" engines of 1010 h.p.; for all a/c with Canadian Airworthiness Certificate at 19,900 lbs. gross wt.

#2-575 (45)

North American "Texan" AT-6 series as 2PCLM with 600 h.p. Wasp R-1340 engine; for all AT-6 and SNJ.

#2-576 (11-28-45)

Douglas "Dragon" B-23 as 2-5PCLM with two 1600 h.p. Wright R-2600 engines; for all USAAF a/c at 26,000 to 27,500 lbs. gross wt.

#2-577 (3-31-47)

Douglas B-18 "Bolo" as 9PCLM with two 1200 h.p. Wright R-1820-53 engines; for all USAAF a/c at 23,200 lbs. gross wt. B-18 Spl. at 21,000 lbs. gross wt. Also B-18A and B-18B.

#2-578 (46)

Noorduyn "Norseman" UC-64 as 4PCL/SM with 600 h.p. Wasp R-1340-AN engine; for all AAF ser. nos. at 7400 lbs. gross wt. 759 built by Canadian Car & Foundry.

#2-579

Taylorcraft TG-6 (glider) as 2PCLM with 65 h.p. Continental A-65 engine; conversion by John Grosse.

#2-580 (11-15-45)

Stearman-Jensen "Ariel" B as 2PCLM with 75 h.p. Continental A-75-9J engine; for one airplane only. Mfgd. by Glenn Stearman.

#2-589

Taylorcraft TG-6 (glider) as 2PCLM with 65 h.p. Continental A-65 engine; conversion by Earl G. Gross.

#2-590

Taylorcraft TG-6 (glider) as 2PCLM with 65 h.p. Continental A-65 engine; conversion by R. V. Black.

#2-591

Taylorcraft TG-6 (glider) as 2PCLM with 65 h.p. Continental A-65 engine; conversion by Midwest Aircraft.

#2-592

Taylorcraft TG-6 (glider) as 2PCLM with 65 h.p. Continental A-65 engine; conversion by Harold L. Barlow.

#2-593

Taylorcraft TG-6 (glider) as 2PCLM with 65 h.p. Continental A-65 engine; conversion by Don F. Swanson.

#2-594

Taylorcraft TG-6 (glider) as 2PCLM with 65 h.p. Continental A-65 engine; conversion by Louis W. Watson.

#2-595

Taylorcraft TG-6 (glider) as 2PCLM with 65 h.p. Continental A-65 engine; conversion by Russell R. Carrington.

#2-596

No information; probably a TG-6 conversion.

#2-597

Taylorcraft TG-6 (glider) as 2PCLM with 65 h.p. Continental A-65 engine; conversion by Union Air Service.

#2-598

Taylorcraft TG-6 (glider) as 2PCLM with 65 h.p. Continental A-65 engine; conversion by C. P. Grace.

#2-599 (9-12-49)

Fairchild M-84-C as 2-4PCLM with 220 h.p. Continental W-670-6N engine; for ser. #6200 only. Probably had PT-19 outer wing panels.

#2-581

Taylorcraft TG-6 (glider) as 2PCLM with 65 h.p. Continental A-65 engine; conversion by McKellar.

#2-582 (5-2-46)

Beech "Kansan" AT-11 and SNB-1 as 4PCLM with two 450 h.p. Wasp R-985 SB engines; for all USAAF & USN ser. nos. at 7835 lbs. gross wt.

#2-583 (7-3-46)

American Eaglecraft "Eaglet" A-31-1B as 2PO/CLM with 50 h.p. Continental A-50-8 or -8F engine; for ser. #1201 and up at 1100 lbs. gross wt. See ATC #450 (Vol. 5-147) for photo.

#2-584 (5-23-46)

Superior (Culver) PQ-8A and TDC-2 as 2PCLM with 125 h.p. Lycoming O-290-B engine; for ser. #L-304-AR thru L-703-AR at 1535 lbs. gross wt. Remodeled by C. M. Jamieson. These were formerly radio-controlled drones.

#2-585

Taylorcraft TG-6 (glider) as 2PCLM with 65 h.p. Continental A-65 engine; conversion by W. C. King.

#2-586

Taylorcraft TG-6 (glider) as 2PCLM with 65 h.p. Continental A-65 engine; conversion by Toth Flying School.

#2-587

Taylorcraft TG-6 (glider) as 2PCLM with 65 h.p. Continental A-65 engine; conversion by Moberly Flying Service.

#2-588

Taylorcraft TG-6 (glider) as 2PCLM with 65 h.p. Continental A-65 engine; conversion by Flight Corp.

#2-600

Taylorcraft TG-6 (glider) as 2PCLM with 65 h.p. Continental A-65 engine; conversion by A. M. Sampson.

#2-601

Taylorcraft TG-6 (glider) as 2PCLM with 65 h.p. Continental A-65 engine; conversion by V. C. Johnson.

#2-602

Taylorcraft TG-6 (glider) as 2PCLM with 65 h.p. Continental A-65 engine; conversion by Edwin Littke.

#2-603

Taylorcraft TG-6 (glider) as 2PCLM with 65 h.p. Continental A-65 engine; conversion by Wm. B. Eaton.

#2-604

Taylorcraft TG-6 (glider) as 2PCLM with 65 h.p. Continental A-65 engine; conversion by Charles Klessig.

#2-605

Taylorcraft TG-6 (glider) as 2PCLM with 65 h.p. Continental A-65 engine; conversion by Salina Aircraft.

#2-606

Taylorcraft TG-6 (glider) as 2PCLM with 65 h.p. Continental A-65 engine; conversion by Leon G. Conover.

#2-607

Taylorcraft TG-6 (glider) as 2PCLM with 65 h.p. Continental A-65 engine; conversion by Clark Hendrickson.

#2-608

Taylorcraft TG-6 (glider) as 2PCLM with 65 h.p. Continental A-65 engine; conversion by T. Cates.

#2-609

Taylorcraft TG-6 (glider) as 2PCLM with 65 h.p. Continental A-65 engine; conversion by R. F. Woolaway.

This photo is typical of the 3-place TG-6 glider conversion into a two-place airplane. It was an easy conversion and the TG-6 was cheap and plentiful; perhaps that is why it was so popular as a modification project.

PHOTO CREDITS
FOR
GROUP 2 SECTION

Aircraft Industries, 130
Alexander Film Co., 105, 124
American Airlines, 110
Balzer, Gerald Collection, 107, 112, 123, 141, 144, 145, 151, 155, 163, 164, 165, 168, 174, 175, 180, 183, 185, 188, 196
Besecker, Roger, 164
Boeing Co., 8, 14, 114, 130, 156, 173, 182, 192
Boeing-Wichita Div., 191, 194
Bowers, Peter Collection, 110, 115, 121, 126, 128, 137, 147, 154, 160, 161, 167, 169, 170, 171, 172, 176, 177, 179, 184, 196, 197, 198, 199
Cessna Aircraft Co., 107, 173, 184
Child, H. Lloyd, 133
Clark, Wm. E., 190
Creswell Photo, 115, 116, 193
Cull, George E., 142
Daniels, John L., 193
Denkelberg Photo, 154
D'Estout, Henri, 169
Douglas Aircraft Co., 186
Fairchild Hiller Corp., 113, 151, 159
Fairchild Republic Co., 196, 203
Ford Motor Co., 186
Grand Central Air Terminal, 138
Havelaar, Marion, 199
Hirsch, Robt. S. Collection, 105, 178
Hoit, R. B., 133
Hudek Collection, 106, 109, 114, 120, 135, 143, 162, 172
Keystone Aircraft Co., 122
Larkins, Wm. T., 158, 166, 191, 200, 201, 202, 204, 205
Linburg Studio, 160
Macdonald Photo, 125, 160
Mayborn, Mitch Collection, 117, 195
McVickar, F. C., 138, 175
Meyers, Chas. W., 109
Northrop Corp., 163, 171, 187
Northwest Airlines
Oberg, Roy, 142
Pan American World Airways, 148
Payette Collection, 122, 139, 150
Poiron, Gene, 113
Pratt & Whitney Div., 111, 149, 155
Reed, Boardman C., 165, 167
Reid Studio, 106
Robbins Studio, 126
Scheetz, Charles, 131, 140
Shaw, Wm. U., 176
Sikorsky Aircraft Div., 112, 119
Smith, Edgar B., 134
Smithsonian Institution, 118, 150, 152, 156, 157, 161, 189
Spartan Aircraft Co., 173
Spencer & Wyckoff, 174
Stearman Aircraft Co., 116
Superior Studio, 202
Steele, Bob Collection, 121
Thompson Photo, 158
Underwood, John W. Collection, 141, 146
United Aircraft Corp., 195
Van Rossem Photo, 153, 159
Weaver, Truman Collection, 127
Whittington, Dick Photo, 181
Williams, Gordon S., 124, 136, 157, 161, 166, 169, 180
Wilson, Kenneth D. Collection, 144, 187
Winstead Photo, 108

THE
LIMITED TYPE CERTIFICATE
(LTC)
FOR
MILITARY AIRPLANES
CONVERTED TO CIVIL USE

LTC-1-3 (12-2-46)

Boeing "Flying Fortress" B-17F and B-17G with four 1000 h.p. Wright R-1820-97 engines at 59,000 lbs. gross wt. Some 18 of the B-17F/B-17G were active at one time as crop-dusters & firefighters; a few are still being used. Approval held by Transcontinental & Western Air, Inc. at Washington, D.C.

LTC-2-3 (12-6-46)

North American "Mitchell" B-25C, B-25G, -25H, -25J, & RB-25 with two 1500-1700 h.p. Wright R-2600-13, -29 engines at 34,000 lbs. gross wt. Approval issued 8-11-58 for B-25N & TB-25N with Wright R-2600-29A & -35 engines. Approval held by Shell Aviation Corp. at New York, N.Y.

LTC-3-3 (12-19-46)

Douglas "Invader" A-26B, -26C with two 1600-2000 h.p. P&W R-2800-21, -27, -41, -71, -79 engines at 35,000 lbs. gross wt. Some 2450 of A-26 were delivered; became B-26 in redesignation of military types. Approval held by Paul V. Shields at New York, N.Y.

LTC-4-3 (1-9-47)

Douglas "Dauntless" A-24B (USAAF) & SBD-5 (Navy) with 1100-1200 h.p. Wright R-1820-60 engine at 9000 lbs. gross wt. Some 5936 delivered to Navy as SBD & 863 delivered to USAAF as A-24. Approval held by Seaboard & Western Airlines at New York, N.Y.

LTC-5-2 (2-19-47)

Convair PB2Y-3, -3R, -5, -5R, -5Z with two 1100 h.p. P&W R-1830-88, -92, -94, -82 engines at 66,000 lbs. gross wt. Approval held by Robert M. Lewis at San Francisco, Calif.

LTC-6-3 (2-21-47)

Convair (Army) LB-30 with four 1200 h.p. P&W (S3C4G) R-1830-33, -67 engines at 58,000 lbs. gross wt. Minimum crew of 3 is required. Approval held by Consolidated-Vultee Aircraft Corp. at San Diego, Calif.

LTC-7-2 (2-28-47)

Sikorsky R-4B with 185-200 h.p. Warner R-550-1, -3 engine at 2700 lbs. gross wt. Approval held by Douglas W. Holmes at Compton, Calif.

LTC-8-2 (3-17-47)

Grumman "Avenger" TBF-1, -1C, TBM-1, -1C, TBM-3, -3E with 1500-1900 h.p. Wright R-2600-8 engine at 17,600 lbs. gross wt. 2293 of TBF by Grumman & 7546 of TBM by Eastern Aircraft. Approval held by Air Trading Corp. at New York, N.Y.

LTC-9-2 (3-31-47)

Douglas "Havoc" A-20B, C, G, H, J with two 1200-1600 h.p. Wright R-2600-11, -23, -29 engines at 25,200 lbs. gross wt. Nearly 6300 were delivered. Approval held by Douglas Aircraft Co.

LTC-10-3 (4-7-47)

Lockheed "Lightning" P-38E, J (F5E), L (F5F/G), P-38M with two 1000-1425 h.p. Allison V-1710-27, -29, -89, -91 engines at 18,500 lbs. gross wt. Some 9924 were built for WW2. Approval held by Lockheed Aircraft Corp. at Burbank, Calif.

LTC-11-2 (4-10-47)

North American "Mustang" P-51C, D, E, K with 1000-1490 h.p. Rolls-Royce V-1650-3 engine at 10,500 lbs. gross wt. Over 15,575 built for WW2. Approval held by DePonti Aviation Co. at Minneapolis, Minn.

LTC-12-2 (4-14-47)

Beech AT-10, -10BH, -10GH, -10GF with two 300 h.p. Lycoming R-680-9, -13 engines at 6600 lbs. gross wt. Cannot be used to haul passengers or cargo for hire. AT-10 was a plywood edition of Beech 18, but had a single tail. Beech built 1771 & Globe (Temco) built 600. Approval first issued 11-21-46 & held by Leland H. Cameron at Chicago, Ill.

LTC-13-2 (4-15-47)

Lockheed "Ventura" B-34, PV-1, PV-2 with two P & W 1600-2000 h.p. R-2800-31 engines at 26,500 lbs. gross wt. Development of "Lodestar"; PV-1 as patrol-bomber & PV-2 as torpedo-bomber. 893 as "Ventura", 1600 as PV-1 & 535 as PV-2. Approval held by Air Trading Corp. at New York, N.Y.

LTC-14-2 (4-23-47)

Northrop "Black Widow" P-61, A, B with two 1600-2000 h.p. R-2800-10, -10W, -65 engines at 30,000 lbs. gross wt. Some 706 of "Black Widow" were built. Approval held by Northrop Aircraft Corp. at Hawthorne, Calif.

LTC-15-2 (4-30-47)

North American "Apache" A-36A with 1000-1325 h.p. Allison V-1710-87 engine at 10,700 lbs. gross wt. "Attack" airplane similar to P-51 "Mustang". Approval held by Woodrow W. "Woody" Edmondson at Lynchburg, Va.

LTC-16-2 (5-6-47)

Curtiss "Owl" 0-52 with 600 h.p. P & W R-1340-51 engine at 5360 lbs. gross wt. "Owl" was active prior to WW2 for observation/photo. Approval held by Holmberg Aerial Survey Co. at Washington, D.C.

LTC-17-2 (5-6-47)

Grumman "Duck" J2F-3, -4, -5, -6 with 900-1050 h.p. Wright R-1820-30, -34, -50, -54 engine at 6530 lbs. gross wt.; 7185 lbs. allowed with 1200-1325 h.p. engines. 650 built. Approval held by R. B. Utterback at Long Beach, Calif.

LTC-18-2 (5-8-47)

Curtiss P-40L, N with 1200-1325 h.p. engines; V-1650-1 & V-1710-73 for P-40L & V-1710-81, -99, -115 for P-40N at 8260 lbs. gross wt. Approval held by Boardman C. Reed at Pasadena, Calif.

LTC-19-2 (5-15-47)

Sikorsky (Helicopter) R-5A with 450 h.p. P & W R-985-AN7 engine at 5380 lbs. gross wt. Approval held by Hel-I-Cop Advertising Corp. at Pasadena, Calif.

LTC-20-2 (6-6-47)

Martin "Mariner" PBM-5 with two 1700-2100 h.p. R-2800-22, -34 engines at 50,000 lbs. gross wt. Approval held by C. F. Krogmann at Wash., D.C.

LTC-21-3 (7-17-47)

Bell P-63C, -63E with 1050 h.p. engines; V-1710-117 for -63C & V-1710-109 for -63E at 11,000 lbs. gross wt. Bell delivered 3300 a/c, 2/3 of which were lend-lease to Russia. In 1946-47 four a/c were stripped for racing. Approval held by Bell Aircraft Corp. at Buffalo, N.Y.

LTC-22-2 (8-8-47)

North American BC-1 with 600 h.p. P & W. R-1340-AN1, -47, -49 engines at 5625 lbs. gross wt. An early variant of AT-6 series. Approval held by Mustang Aviation, Inc. at Dallas, Tex.

LTC-23-2 (8-29-47)

Grumman "Bearcat" F8F-1 with 1700-2100 h.p. P & W R-2800-34, -34W engines at 9000 lbs. gross wt. Some 1265 were delivered. Approval held by the Grumman Aircraft Engrg. at Bethpage, LI.,N.Y.

LTC-24-2 (10-8-47)

Vought OS2U-1, -2, -3, & OS2N-1 with 450 h.p. P & W R-985-AN2, -48, -50 engines at 5350 lbs. gross wt.; 5200 lbs. for rough water operation. Approval held by Carl F. Krogmann at Washington, D.C.

LTC-25-2 (11-5-47)

Grumman "Wildcat" FM-2 with 1200-1350 h.p. Wright R-1820-56 engine at 7500 lbs. gross wt.; a GM-built version of XF4F-8. GM built over 5900 of FM-2. Approval held by Richard R. Carlisle at Tarrant, Ala.

LTC-26-2 (11-17-47)

Stinson "Vigilant" L-1, A, B, C, D, E, F with 300 h.p. Lycoming R-680-90, -13 engines at 3325-3385 lbs. gross wt. First as 0-49; 324 were delivered in 1940-41. Approval held by Executive Airlines, Inc. at Cleveland, Ohio.

LTC-27-2 (11-18-47)

North American "Basic Trainer" BT-9, A, B, C with 400 h.p. Wright R-975-7 engine at 4500 lbs. gross wt. Approval held by James O. Wyatt at Dyersburg, Tenn.

LTC-28-2 (12-16-47)

Superior (Culver) PQ-14A, B, YPQ-14A, B, and TD2C-1 with 130-155 h.p. Franklin 0-300-11 engine at 1820 lbs. gross wt. Some 2040 were built as radio-controlled drones. Approval held by N. A. Kalt at Dallas, Tex.

LTC-29-1 (1-5-48)

Sikorsky (Helicopter) YR-6A, R-6A, HOS-1 with 235 h.p. Franklin 0-405-9 engine at 2700 lbs. gross wt. Approval held by Stolp-Adams Co. at Compton, Calif.

LTC-30-2 (1-6-48)

Convair (Army) C-87A with four 1200 h.p. P. & W R-1830-43, -65 engines at 56,000 lbs. gross wt. A transport variant of the B-24 bomber. Some 282 were built. Approval held by Wm. P. Odom at Roslyn, L.I., N.Y.

LTC-31-2 (1-14-48)

Curtiss "Flying Jeep" AT-9, -9A with two 300 h.p. Lycoming R-680-9 engines at 6000 lbs. gross wt. A transition trainer. Approval held by L. S. Rehr at Coral Gables, Fla.

LTC-32-1 (1-14-48)

North American "Basic Trainer" BT-14 with 450 h.p. P & W R-985-25 engine at 4470 lbs. gross wt. Approval held by P.J. Franklin at Culver City, Calif.

LTC-33-1 (Petition date 1-14-52; Specification issued 7-14-59)

Martin "Marauder" B-26C with two 1700-2100 h.p. P & W R-2800-83AM8 or R-2800-83AM11 Engines. Over 4700 of B-26 built for WW2. Approval held by Tenn. Gas Transmission Co. at Houston, Tex.

This listing provided by courtesy of Wm. T. Larkins.

THE
RESTRICTED CATEGORY
(AR)
FOR AIRPLANES CONVERTED
TO RESTRICTED USE

AR-1

Curtiss "Warhawk" P-40 series (all models except P-40L and P-40N) with Allison V-1710 & Rolls-Royce (Packard "Merlin") V-1650 engines. Approval held by Rogue River Valley Traffic Assoc. at Medford, Ore.

AR-2

Naval Aircraft Factory N3N-1 "Primary Trainer" as 1POLB with Wright R-760 engines. Converted to dusters & sprayers. Approval held by Crowl Dusters, Inc. at Phoenix, Ariz.

AR-3

Federal "Primary Trainer" XPT-1 as 2POLB with Wright R-760 engine. Only 2 built, had all-metal fuselage. Approval held by Wilson Air Service at Bridgeton, N.Y.

AR-4

DeHaviland DH-98 "Mosquito" (USAAF as F-8) with two Packard "Merlin" V-1650 engines at 21,000 lbs. gross wt. U.S. production was contemplated, but not undertaken. Designed to do photo-recon. Approval held by Hurd Mapping Co. of Minn.; approval was cancelled 4-15-55 by holder's request. Refer to AR-13.

AR-5

Avro "Anson" series V twin-engined trainer (British). Approval held by Continental Oil Co. at Ponca City, Okla.

AR-6

St. Louis "Primary Trainer" YPT-15 as 2POLB with Wright R-760 engines at 2770 lbs. gross wt.; about 14 were delivered to USAAF in 1940. Approval held by Arizona Aviation Service at Safford, Ariz.

AR-7

Piper "Super Cub" PA-18A with 125 h.p. Lycoming 0-290-D engine, approved 2-5-52. The PA-18A-135 with Lycoming 0-290-D2 engine was approved 6-17-52. The PA-18A-150 with Lycoming 0-320-A2A or -A2B engine was approved 10-7-54; all were as 1PCLM. Approval probably held by Piper Aircraft Corp. at Lock Haven, Pa.

AR-8

Douglas "Dauntless" A-24A (Army) as 1PCLM converted to pest-control & mosquito-sprayer. In U.S. Navy as SBD. Approval held by City of Portland at Portland, Ore.

AR-9

Stearman "Primary Trainer" Model 73 (Navy NS-1) with 220 h.p. Wright J5 engine. As 2POLB; prototype for PT-13/17 & N2S. Approval held by Southwest Aircraft, Inc. at Ft. Worth, Tex.

AR-10

Convair L-13 series as 2PCLM with 245 h.p. Franklin 0-425-9 engine; adaptable to various conversions. A development of the Stinson L-5 "Sentinel". Approval held by Lester Branchflower at Burbank, Calif.

AR-11

Noorduyn "Harvard" (Army AT-16) same as AT-6A; Canadian-built with 600 h.p. R-1340-AN1 engine; 1500 built for RCAF. Approval held by Autair, Ltd. at London, England.

AR-12

Northrop "Reporter" (RF-15) and "Black Widow" (RF-61C) with two 2100 h.p. R-2800 engines. Approval held by Steward-Davis Co. at Gardena, Calif.

AR-13

DeHaviland DH-98 "Mosquito" (USAF F-8) with two Packard "Merlin" (Rolls-Royce) V-1650 engines. Approval held by Trans-World Engrg. Corp. at Los Angeles, Calif.

AR-14

Rausch "Model 18" (type unknown) rumored as racing airplane. Approval held by Wm. L. Rausch at Hackensack, N.J.

AR-15

Fairchild "Packet" C-82 with two R-2800 engines at 42,000 lbs. gross wt. Approval held by Steward-Davis Co. at Gardena, Calif. Some converted to fire-fighters.

AR-16

Boeing-Wichita YL-15G as 2PCLM with 125 h.p. Lycoming 0-290-7 engine; used in Army Ground Force cooperation. Approval held by Norman E. Brunquist at Spenard, Alaska.

AR-17

Piasecki (Helicopter) HRP-1 twin-rotor with 500 h.p. Continental R-9 engine; formerly used by U.S. Navy. Approval held by Alaska Helicopters, Inc. at San Francisco, Calif.

AR-18

Grumman "Hellcat" F6F carrier-fighter with 2000 h.p. R-2800 engine; Grumman built 12,275 for WW2. Approval held by Thomas Jennings at Washington, D.C.

AR-19

Curtiss "Helldiver" SB2C; some 6130 were built for WW2, including 900 with R-2600 engines for USAAF as A-25A. Approval held by Clayton V. Curtis at Visalia,Calif.

AR-20

Temco "Buckaroo" (?) T-35 as 2PCLM primary trainer. Approval held by Wm. E. Quinn at Elko, Nev.

AR-21

Piasecki (Helicopter) HRP-2 as tandem-rotor, all-metal version of the HRP-1. Approval held by Alaska Helicopters, Inc. at San Francisco, Calif.

AR-22

Consolidated-Vultee PBY-6A patrol boat. Approval held by The Babb Co. at Phoenix, Ariz.

AR-23

Seversky P-35 a pre-WW2 fighter with P & W R-1830 engine at 5600 lbs. gross wt.; progenitor of "Thunderbolt". Approval held by Chas. P. Doyle at Rosemount, Minn.

AR-24

Boeing "Stratofreighter" YC-97 with four 3000 h.p. Wasp-Major engines at 130,000 lbs. gross wt.; a military "377". Approval held by DeLong Corp. at New York, N.Y.

AR-25

Chase "Avitruk" YC-122C with two R-2800 engines at 40,000 lbs. gross wt.; designed for assault-cargo. Approval held by Roberts Aircraft Co. at Boise, Idaho.

AR-26

Columbia "Amphibian" XJL-1 (U.S. Navy). Approval held by G. R. Board at Michigan City, Ind.

AR-27

Kaman (Helicopter) K-600 as U.S. Navy HOK-1. Approval held by Kaman Aircraft Corp. at Bloomfield, Conn.

AR-28

Grumman "Tigercat" F7F-3 as twin-engined fighter (U.S. Navy). Some 364 built for use by Marines. A development of the earlier XF5F-1 "Skyrocket". Approval held by Geo. F. Kreitzberg at Salem. Ore.

AR-29

Convair "Privateer" P4Y-2 a Naval version of B-24J with a longer fuselage & a single tail. Some 742 built for U.S. Navy. Several converted for fire-bombing. Approval held by Transaire Spraying Co. at Canyon, Tex.

AR-30

North American T-28 with 800 h.p. Wright R-1300-1A engine. Was popular as sportplane before the "gasoline crunch". Approval issued 9-8-58 & held by Thompson Aircraft Sales at Phoenix, Ariz.

AR-31

No information available.

AR-32

Grumman "Bearcat" F8F-2 with R-2800-30W engine. Approval issued 11-2-59 & held by Acme Aircraft Co. at Lomita, Calif.

AR-33

DeHaviland-Canada "Beaver" L-20A issued 4-1-60 with R-985-AN engines for spl. purpose of search, rescue, & pest control. May be Converted to DHC-2 with DHC modification specs. Approval held by Civil Air Patrol, Inc. at Anchorage, Alaska.

This listing provided by courtesy of Wm. T. Larkins.

MASTER INDEX
FOR
AIRPLANES

A cross-reference Index listing company name, model name, model number, and nickname. Engines used are shown in parentheses with the basic airplane description following. P=Place, O=Open, C=Closed, L=Land, S=Sea, M=Monoplane, B=Biplane, Am=Amphibian, FB=Flying boat, and Ag=Autogiro. Figures at extreme right listed to locate Vol. number and page number. This listing pertains to all 9 vols. of U. S. CIVIL AIRCRAFT.

MASTER INDEX

INDEX
FOR THE PEOPLE OF AVIATION

This is not just an ordinary list of names, it is literally a "who's who" of aviation in its earlier years. Listed here are names that wrought the fiber of aviation for nearly four decades, a vibrant "game" that took all kinds of followers to make it work. Listed are the inventors and designers that had the dreams, the fearless pilots that flew these contrivances sometimes to glory and sometimes to death. Then, there are too the tireless recordbreakers that spanned oceans, continents, or flew against time to go somewhere faster and faster; others flew higher and higher while others flew farther and farther. Among this close-knit clan were the unsung heroes, the craftsmen that shaped these "ships" into being from metal, wood, and cloth, and "angels" and investors that provided the money, and the "company officers" that tried their best to keep the company in business. The aura of aviation also attracted the so-called "sportsman" who came from all walks of life and was usually "flush" enough to buy the finest creations that minds labored to create. Women were making their mark in aviation too, most all of the "barnstormers"now had regular jobs, and the young daredevils had grown up to fly huge airliners across our nation, or spanning oceans to distant playgrounds. Aviation had come a long way in the four decades depicted in this series of books, but it took all these people to make it come about. The mind of man can create wonderful things, who knows what is yet in store.

Jos. P. Juptner

The listing following the name is first the Vol. number and then the page number. This symbol (†) denotes that a photograph is shown.

—A—

—B—

—C—

—I—

—J—

—K—

—Y—

—Z—